Recent Titles in This Series

43 **James E. Humphreys,** Conjugacy classes in semisimple algebraic groups, 1995
42 **Ralph Freese, Jaroslav Ježek, and J. B. Nation,** Free lattices, 1995
41 **Hal L. Smith,** Monotone dynamical systems: an introduction to the theory of competitive and cooperative systems, 1995
40.1 **Daniel Gorenstein, Richard Lyons, and Ronald Solomon,** The classification of the finite simple groups, 1994
39 **Sigurdur Helgason,** Geometric analysis on symmetric spaces, 1993
38 **Guy David and Stephen Semmes,** Analysis of and on uniformly rectifiable sets, 1993
37 **Leonard Lewin, Editor,** Structural properties of polylogarithms, 1991
36 **John B. Conway,** The theory of subnormal operators, 1991
35 **Shreeram S. Abhyankar,** Algebraic geometry for scientists and engineers, 1990
34 **Victor Isakov,** Inverse source problems, 1990
33 **Vladimir G. Berkovich,** Spectral theory and analytic geometry over non-Archimedean fields, 1990
32 **Howard Jacobowitz,** An introduction to CR structures, 1990
31 **Paul J. Sally, Jr. and David A. Vogan, Jr., Editors,** Representation theory and harmonic analysis on semisimple Lie groups, 1989
30 **Thomas W. Cusick and Mary E. Flahive,** The Markoff and Lagrange spectra, 1989
29 **Alan L. T. Paterson,** Amenability, 1988
28 **Richard Beals, Percy Deift, and Carlos Tomei,** Direct and inverse scattering on the line, 1988
27 **Nathan J. Fine,** Basic hypergeometric series and applications, 1988
26 **Hari Bercovici,** Operator theory and arithmetic in H^∞, 1988
25 **Jack K. Hale,** Asymptotic behavior of dissipative systems, 1988
24 **Lance W. Small, Editor,** Noetherian rings and their applications, 1987
23 **E. H. Rothe,** Introduction to various aspects of degree theory in Banach spaces, 1986
22 **Michael E. Taylor,** Noncommutative harmonic analysis, 1986
21 **Albert Baernstein, David Drasin, Peter Duren, and Albert Marden, Editors,** The Bieberbach conjecture: Proceedings of the symposium on the occasion of the proof, 1986
20 **Kenneth R. Goodearl,** Partially ordered abelian groups with interpolation, 1986
19 **Gregory V. Chudnovsky,** Contributions to the theory of transcendental numbers, 1984
18 **Frank B. Knight,** Essentials of Brownian motion and diffusion, 1981
17 **Le Baron O. Ferguson,** Approximation by polynomials with integral coefficients, 1980
16 **O. Timothy O'Meara,** Symplectic groups, 1978
15 **J. Diestel and J. J. Uhl, Jr.,** Vector measures, 1977
14 **V. Guillemin and S. Sternberg,** Geometric asymptotics, 1977
13 **C. Pearcy, Editor,** Topics in operator theory, 1974
12 **J. R. Isbell,** Uniform spaces, 1964
11 **J. Cronin,** Fixed points and topological degree in nonlinear analysis, 1964
10 **R. Ayoub,** An introduction to the analytic theory of numbers, 1963
9 **Arthur Sard,** Linear approximation, 1963
8 **J. Lehner,** Discontinuous groups and automorphic functions, 1964
7.2 **A. H. Clifford and G. B. Preston,** The algebraic theory of semigroups, Volume II, 1961
7.1 **A. H. Clifford and G. B. Preston,** The algebraic theory of semigroups, Volume I, 1961
6 **C. C. Chevalley,** Introduction to the theory of algebraic functions of one variable, 1951
5 **S. Bergman,** The kernel function and conformal mapping, 1950

(*Continued in the back of this publication*)

Conjugacy Classes in Semisimple Algebraic Groups

MATHEMATICAL
Surveys and Monographs

Volume 43

Conjugacy Classes in Semisimple Algebraic Groups

James E. Humphreys

American Mathematical Society
Providence, Rhode Island

The author was supported in part by the National Science Foundation.

1991 *Mathematics Subject Classification*. Primary 20G15; Secondary 17B20, 22E10.

Library of Congress Cataloging-in-Publication Data

Humphreys, James E.
Conjugacy classes in semisimple algebraic groups / James E. Humphreys.
p. cm. — (Mathematical surveys and monographs; v. 43)
Includes bibliographical references and index.
ISBN 0-8218-0333-6 (alk. paper)
1. Linear algebraic groups. 2. Lie algebras. 3. Semisimple Lie groups. I. Title. II. Series: Mathematical surveys and monographs; no. 43.
QA179.H86 1995
512′.55–dc20 95-16546
CIP

∞ The paper used in this book is acid-free and falls within the guidelines
established to ensure permanence and durability.
♻ Printed on recycled paper.
This volume was printed directly from copy prepared by the author using $\mathcal{A}\!\mathcal{M}\!\mathcal{S}$-LaTeX,
the American Mathematical Society's TeX macro system.

10 9 8 7 6 5 4 3 2 1 00 99 98 97 96 95

To the memory of
Roger Richardson

Contents

Preface

A major landmark in the theory of linear algebraic groups was Chevalley's classification of simple algebraic groups over an arbitrary algebraically closed field in the late 1950s. This was followed in the 1960s by the work of Borel and Tits on the structure and classification of reductive groups relative to an arbitrary field of definition. In the period since 1965, many deep results have emerged from the systematic study of conjugacy classes and centralizers. Much of this work continues to be applied to the structure and representation theory of finite groups of Lie type, real or p-adic groups, and related Lie algebras. An important subtheme since the mid-1970s has been Springer's cohomological construction of Weyl group representations.

My purpose here is to draw together the most significant results about classes and centralizers in a connected semisimple group, working as a rule over an algebraically closed field K of arbitrary characteristic. While the natural setting is that of reductive groups such as $\mathrm{GL}(n, K)$, one can avoid distracting notational complications by working mainly with semisimple groups such as $\mathrm{SL}(n, K)$. Generalizations to the reductive case are almost always effortless, as long as the groups are connected.

Here are some of the high points in the development, with G denoting a connected semisimple (or reductive) group:

- Steinberg's theorem on connectedness of centralizers of semisimple elements when G is simply connected.
- Steinberg's description of the algebra of class functions on G.
- Theorems of Richardson and Lusztig on the finiteness of the number of unipotent classes in G.
- Steinberg's characterizations of regular elements.
- Richardson's dense orbit theorem for parabolic subgroups, and the more general Lusztig–Spaltenstein theory of induction of unipotent classes.
- Springer's resolution of singularities for the unipotent variety.
- Dimension formula of Springer–Steinberg for fixed point sets on the flag variety, and equidimensionality theorem of Spaltenstein for the irreducible components of such sets.

- Description of Dynkin curves related to subregular unipotent elements.
- Springer's identification of the unipotent and nilpotent varieties in good characteristic.
- Parametrization of unipotent classes by weighted Dynkin diagrams in good characteristic (Dynkin, Kostant, Bala–Carter, Pommerening).
- Applications to finite groups of Lie type.
- Springer's representations of Weyl groups.

This book is both a monograph and a survey. In the first six chapters, complete proofs are given for the main results. As in the texts [**15, 84, 177**], group-theoretic reasoning (sometimes intricate) is combined with relatively elementary arguments from algebraic geometry. But more sophisticated ideas from algebraic geometry begin to show up toward the end of Chapter 6, for example in 6.20. These chapters progress from the most general and accessible results to the more complicated ones, which may require case-by-case study or avoidance of "bad" prime characteristics.

Chapters 7–10 provide detailed surveys, with examples and indications of proof as well as reasonably complete references to the literature (including some related papers not discussed directly here). But I have tried to avoid too much overlap with the books of Carter [**29**] and Collingwood–McGovern [**31**].

The reader is expected to be familiar with the structure and classification of semisimple groups over an algebraically closed field K of arbitrary characteristic. Here the basic vocabulary includes: maximal tori, Borel and parabolic subgroups, Weyl group, Bruhat decomposition, Lie algebra, root system. For convenience, the main definitions and theorems are summarized in Chapter 0. While $\mathrm{SL}(n, K)$ may serve as a ready example, the behavior of its conjugacy classes is often far nicer than in other cases. For this reason the less tidy behavior of the exceptional group of type G_2 is invoked from time to time, especially to illustrate theorems about unipotent classes.

My interest in this subject matter goes back to my graduate studies at Yale in 1963–66, when I was introduced to the Chevalley seminar as well as Steinberg's paper [**184**] on regular elements. Giving talks on that paper was a challenging (but rewarding) assignment for an unsuspecting graduate student. My aim here is to give an updated and more comprehensive exposition of the themes in Steinberg's 1972 Tata lectures [**188**]. Although the subject may now be regarded as mature, it remains a challenge to find more direct or unified proofs of some theorems—avoiding case-by-case verifications, for example, when char K is small.

I am grateful to the many colleagues who have provided advice along the way, and to the National Science Foundation for research support. While I have tried to attribute results correctly, the complicated evolution of some of the ideas may have resulted in some oversights. Comments and corrections from readers will be welcome.

Frequently Used Symbols

The symbols listed here occur consistently throughout the book, while others may vary locally. As a general convention, we write $A \subset B$ to mean that A is a subset of B, with $A = B$ permitted. We use $\mathbb{Z}, \mathbb{Q}, \mathbb{R}, \mathbb{C}, \mathbb{F}_q$ to denote (respectively) integers, rational numbers, real numbers, complex numbers, and a finite field of order q.

K	algebraically closed field
G	connected semisimple algebraic group
e	identity element of group
r	rank of G
$Z(G)$	center of G
$\mathrm{Cl}(x)$	conjugacy class of $x \in G$
$C_G(x)$	centralizer of element $x \in G$
$\mathfrak{g}$	Lie algebra of G
$\mathrm{Ad}\, x$	adjoint operation by $x \in G$ on $\mathfrak{g}$
T	maximal torus of G
N	normalizer $N_G(T)$
W	Weyl group N/T of G relative to T
Φ	root system of G relative to T
Δ	system of simple roots $\{\alpha_1, \ldots, \alpha_r\} \subset \Phi$
$X(T)$	character group $\mathrm{Hom}(T, K^\times)$
$Y(T)$	cocharacter group $\mathrm{Hom}(K^\times, T)$
U_α	one-dimensional root subgroup $\{x_\alpha(c) \mid c \in K\}$
B	Borel subgroup of G
U	unipotent radical of Borel subgroup
P	parabolic subgroup of G
L	Levi subgroup of parabolic subgroup
V	unipotent radical of parabolic subgroup
$\mathcal{B}$	flag variety of G (set of Borel subgroups)
$\mathcal{U}$	unipotent variety of G
$\mathcal{N}$	nilpotent variety of $\mathfrak{g}$

Dynkin Diagrams

Here are the Dynkin diagrams of the various irreducible root systems, with vertices numbered as in Bourbaki [**23**, Planches]. Care must be taken in consulting other books, where different numbering systems or orientations of the diagrams are sometimes found. For example, Carter [**29**, 13.1] orients the diagrams of types G_2, E_7, and E_8 in the reverse sense.

$A_r (r \geq 1)$:

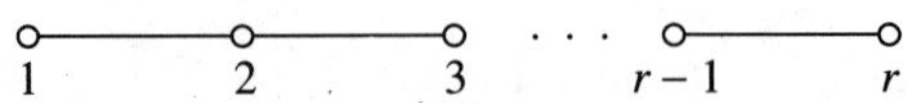

$B_r (r \geq 2)$:

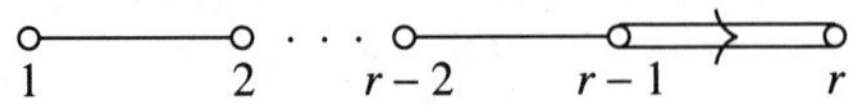

$C_r (r \geq 3)$:

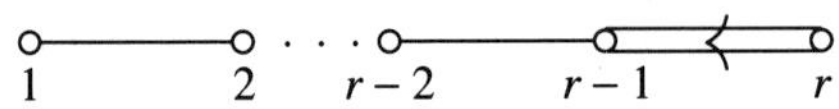

$D_r (r \geq 4)$:

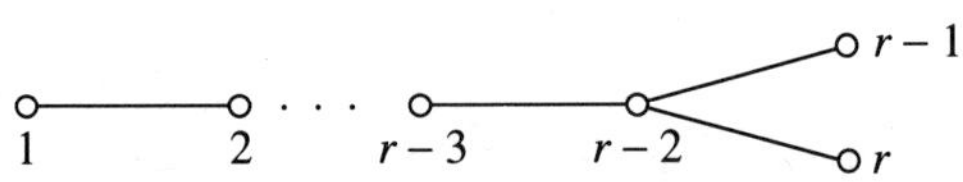

E_6:

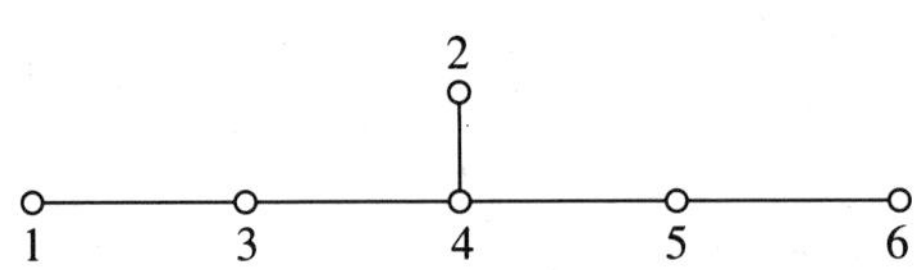

E_7:

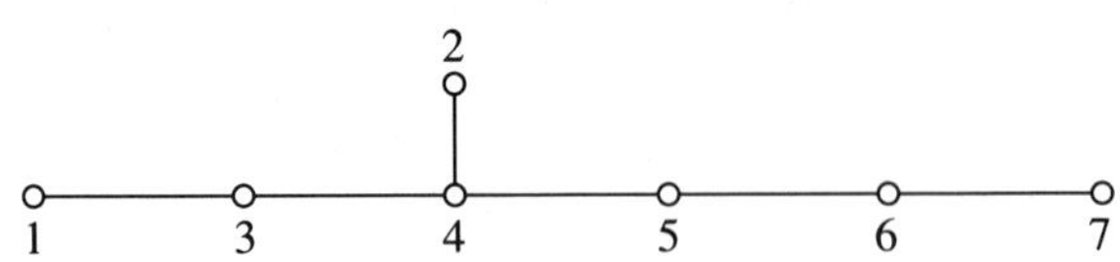

E_8:

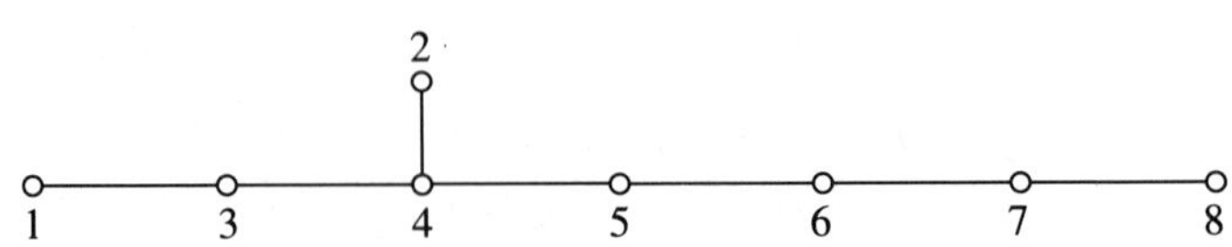

F_4:

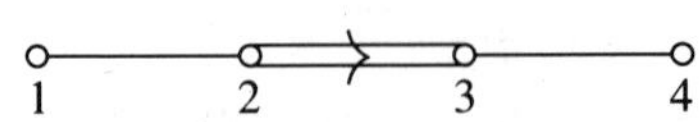

G_2:

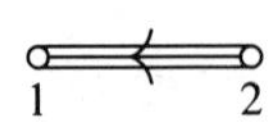

CHAPTER 0

Review of Semisimple Groups

In this chapter we assemble some basic facts about the structure of semisimple and reductive groups, with references (abbreviated using the respective letters B, H, S) to the textbooks by Borel [**15**], Humphreys [**84**], Springer [**177**]. For example, a reference to section 2.3 of Borel will be given as B2.3. Note that the summaries below do not reflect the order in which results are usually derived.

Starting with 0.3, the letter G will be reserved for a connected semisimple group over an algebraically closed field K. The identity element of a group is always denoted e, except for matrix groups such as $\mathrm{GL}(n, K)$—in which case the symbol 1 is used. The Lie algebra of an algebraic group G is denoted by $\operatorname{Lie} G$ or $\mathfrak{g}$. Lower case letters are used for elements of groups or Lie algebras.

0.1. Jordan–Chevalley Decomposition in Algebraic Groups

This is a fundamental tool in the study of any (linear) algebraic group G. Each element x of G has a unique decomposition $x = x_s x_u = x_u x_s$, where x_s is **semisimple** and x_u is **unipotent**: in some (hence any) realization of G as a closed subgroup of $\mathrm{GL}(n, K)$, the matrix x_s is diagonalizable and the matrix x_u has all eigenvalues equal to 1. There is an analogous Jordan–Chevalley decomposition in the Lie algebra $\mathfrak{g}$ of G: each element is uniquely the sum of a semisimple and a nilpotent element whose Lie bracket is 0 (B§4, H§15, S2.4 and S3.4).

A **torus** is a connected algebraic group consisting of semisimple elements; it is automatically commutative, and is in fact isomorphic to a direct product of copies of the multiplicative group $K^\times$ (B8.5, H16.2, S2.5). A **maximal torus** of G is a closed subgroup of largest possible dimension isomorphic to a torus—for example, the group of diagonal matrices in $\mathrm{GL}(n, K)$ or $\mathrm{SL}(n, K)$. When G is connected, each semisimple element of G lies in some maximal torus, while each unipotent element lies in a maximal connected **unipotent subgroup**: a closed subgroup consisting of unipotent elements (B11.10, H22.2, S7.3.3).

There are similar results in $\mathfrak{g}$: an element $x \in \mathfrak{g}$ is semisimple if and only if x lies in $\operatorname{Lie} T$ for some maximal torus T of G (B11.8), while x is nilpotent if and only if it lies in the Lie algebra of a closed unipotent subgroup of G (B14.26).

0.2. Semisimple and Reductive Groups

Any algebraic group G has a unique largest closed connected normal unipotent subgroup $R_u(G)$, the **unipotent radical**, and a unique largest closed connected normal solvable subgroup $R(G)$, the **radical**. If $R_u(G)$ is trivial, G is called **reductive**, while if $R(G)$ is trivial, G is called **semisimple** (B11.21, H19.5, S6.14). Note that, contrary to the convention in H, we do not assume that a semisimple or reductive group is connected. While our main object of study will be a connected semisimple group G, there are naturally occurring reductive subgroups of G which need not be connected: for example, centralizers of semisimple elements.

We say the algebraic group G has a **Levi decomposition** if $G = HR_u(G)$, where H is a reductive subgroup (called a **Levi factor**) intersecting $R_u(G)$ trivially. When $\operatorname{char} K = 0$, Levi factors always exist and are conjugate under $R_u(G)$, but this sometimes fails in prime characteristic.

0.3. Borel Subgroups

Here the structure theory begins in earnest (B§11; H, Chap. VIII; S, Chap. 7). Unless otherwise specified, we assume from now on that *G is a connected semisimple group.* A **Borel subgroup** B of G is a maximal closed connected solvable subgroup. All of these are conjugate in G, and their union is G. The unipotent radical U of B is a maximal connected unipotent subgroup of G. Any maximal torus T of B is also a maximal torus of G, and all maximal tori of G are conjugate; their common dimension r is called the **rank** of G.

If T is a maximal torus of B, then $B = TU$ (a Levi decomposition). Moreover, U is equal to the derived group (B, B), while $\dim U/(U,U) = r = \dim T$. The center $Z(G)$ of G is finite and coincides with $Z(B)$ for any Borel subgroup B. Given B and T, there is a unique **opposite** Borel subgroup $B^- = TU^-$ intersecting B precisely in the common maximal torus T.

The **flag variety** $\mathcal{B}$ is the collection of all Borel subgroups of G. By fixing one Borel subgroup B, and using the fact that $N_G(B) = B$, we may identify $\mathcal{B}$ with the projective variety G/B.

The Lie algebra of a Borel subgroup is called a **Borel subalgebra** of $\mathfrak{g}$, and has a vector space decomposition $\mathfrak{b} = \mathfrak{t} \oplus \mathfrak{u}$, where $\mathfrak{u}$ is an ideal. The union of all Borel subalgebras is $\mathfrak{g}$ (B14.25).

0.4. Roots

Fix a maximal torus T of G. Associated to T are two abelian groups, the **character group** $X(T) := \operatorname{Hom}(T, K^\times)$ and the **cocharacter group** $Y(T) := \operatorname{Hom}(K^\times, T)$. Here $K^\times$ is the multiplicative group of K, viewed as a 1-dimensional torus. In each group, the operation comes from multiplying function values, but it is conventional to write this operation additively. Each group is free abelian of rank $r = \dim T = \operatorname{rank} G$. There is a nondegenerate pairing $X(T) \times Y(T) \to \mathbb{Z}$, sending the pair (χ, λ) to the integer $n = \langle \chi, \lambda \rangle$ which satisfies: $\chi(\lambda(c)) = c^n$ for all $c \in K^\times$.

The adjoint action of T on the Lie algebra $\mathfrak{g}$ of G leads to a root space decomposition. Besides the Lie algebra $\mathfrak{t}$ of T, which is fixed pointwise by $\operatorname{Ad} T$, there are 1-dimensional subspaces of $\mathfrak{g}$ corresponding to nontrivial characters α of T, called **roots** (B13.8, H26.2, S9.3):

$$\mathfrak{u}_\alpha := \{x \in \mathfrak{g} \mid (\operatorname{Ad} t)(x) = \alpha(t)x \text{ for all } t \in T\}.$$

Then $\mathfrak{g} = \mathfrak{t} \oplus \bigoplus \mathfrak{u}_\alpha$, and the adjoint action of $\mathfrak{t}$ on the root space $\mathfrak{u}_\alpha$ is given by the differential $d\alpha$. Denote by $\Phi(T, G)$ (or just Φ) the set of all roots of G relative to T. Since all maximal tori are conjugate, Φ is essentially independent of T and is called the **root system** of G. Root systems may be defined axiomatically as finite sets of vectors in a euclidean space, realized here as $X(T) \otimes_{\mathbb{Z}} \mathbb{R}$ (see B24.7, H, Appendix, as well as Bourbaki [**23**, Chap. VI] and Humphreys [**82**, Chap. III]). The irreducible ones (associated with simple algebraic groups or simple Lie algebras over $\mathbb{C}$) are classified by **Dynkin diagrams**. These are pictured above, with the conventional numbering of vertices.

0.5. Root Groups

To each root α corresponds a unique closed connected 1-dimensional subgroup U_α of G normalized by T having $\mathfrak{u}_\alpha$ as Lie algebra (B13.18, H26.3, H28.1, S9.3.6). We call U_α a **root group**. There is an isomorphism from the additive group of K onto U_α, denoted $c \mapsto x_\alpha(c)$, satisfying $tx_\alpha(c)t^{-1} = x_\alpha(\alpha(t)c)$ for all $t \in T$. If $B = TU$ is a Borel subgroup, then U can be written (in any order) as the product of certain root groups U_α, $\alpha \in \Phi^+$. These α are called **positive roots**. Their inverses (written $-\alpha$ in additive notation) are **negative roots**, and the corresponding root groups occur in the unipotent radical U^- of the opposite Borel subgroup. Positive systems are determined by **simple systems**: a linearly independent set $\Delta \subset \Phi$ consists of simple roots if each positive root is a $\mathbb{Z}^+$-linear combination of Δ. When G is semisimple, $\operatorname{rank} G = \dim T = |\Delta|$.

If $\alpha, \beta \in \Phi^+$, commutators in U of the form $(x_\alpha(c), x_\beta(d))$ lie in the product of root groups U_γ with $\gamma = r\alpha + s\beta$ for integers $r, s > 0$. The structure constants occurring in such commutation formulas can in fact be taken to lie in the prime field of K (B14.5, H32.5, S10.1.4).

The standard rank one subgroups G_α of G corresponding to pairs of roots $\pm\alpha$ (each isomorphic to $\mathrm{SL}(2, K)$ or $\mathrm{PGL}(2, K)$) yield **coroots** $\alpha^\vee \in Y$. Here $\alpha^\vee$ maps $c \in K^*$ to the image in G_α of the matrix $\operatorname{diag}(c, c^{-1}) \in \mathrm{SL}(2, K)$. In particular, $\langle \alpha, \alpha^\vee \rangle = 2$. It is easily seen that the set $\Phi^\vee$ of coroots forms an abstract root system in the Euclidean space obtained from $Y(T)$ by extending scalars to $\mathbb{R}$. Its Weyl group is canonically isomorphic to the Weyl group W of G relative to T.

0.6. Weyl Group

A maximal torus T of G is equal to its centralizer. Its normalizer $N_G(T)$ is often denoted N. Here N/T is the **Weyl group** W of G relative to T. Since all maximal tori are conjugate, the finite group W is an invariant of G up to isomorphism. For example, the Weyl group of $\mathrm{SL}(n, K)$ is the symmetric group S_n. The group W permutes the roots and root groups in a natural way: if $n_w \in N$ represents $w \in W$, then $n_w U_\alpha n_w^{-1} = U_{w(\alpha)}$ (H26.3).

In the setting of abstract root systems, the Weyl group arises as the group generated by reflections s_α relative to roots $\alpha \in \Phi$. To each root α corresponds a **coroot** $\alpha^\vee$ (as above) for which $\langle \alpha, \alpha^\vee \rangle = 2$, with $s_\alpha(\chi) = \chi - \langle \chi, \alpha^\vee \rangle \alpha$. If $\Delta = \{\alpha_1, \ldots, \alpha_r\}$ is a simple system in Φ, then W is already generated by the **simple reflections** s_{α_i}. The smallest number of simple reflections in an expression for w is then called the **length** $\ell(w)$. There is a unique element $w_\circ$ of maximum length (equal to the number of positive roots), and $w_\circ(\Phi^+) = \Phi^-$.

0.7. Bruhat Decomposition

Here we reach the core of the structure theory (B§14, H§28, S10.2).

Fix a Borel subgroup $B = TU$ in G, and write $N = N_G(T)$ and $W = N/T$. Choose a representative $n_w \in N$ for each $w \in W$. When there is no ambiguity, we may write BwB, wB, etc. in place of Bn_wB, n_wB, etc. Similarly, write ${}^wB = n_w B n_w^{-1}$ and ${}^wU = n_w U n_w^{-1}$. The choice of B determines a simple system Δ and positive system Φ^+ in the root system Φ of G relative to T.

Let $w_\circ = w_\circ^{-1}$ be the unique element of W interchanging positive and negative roots, and let $B^- = TU^-$ be the opposite Borel subgroup $w_\circ B w_\circ$. For $w \in W$, define $U'_w = U \cap {}^wU^-$, the product of those U_α for which $\alpha > 0$ but $w^{-1}\alpha < 0$. With this notation, the **Bruhat decomposition** may be formulated as:

$$G = \bigcup_{w \in W} BwB = \bigcup_{w \in W} U'_w wB.$$

Here the double cosets BwB are disjoint for distinct elements of W, and each element of $U'_w wB$ may be written uniquely in the form un_wb (once a representative $n_w \in N$ of w has been fixed).

When $w = w_\circ$, we get an open subset $Bw_\circ B = Uw_\circ B = w_\circ U^- B$; either it or its translate U^-B is often called the **big cell**. The Bruhat decomposition of G yields a cell decomposition: $G/B = \bigcup BwB/B$. The cell BwB/B is isomorphic to an affine space of dimension $\ell(w)$; its closure in G/B is called a **Schubert variety**.

In deriving the Bruhat decomposition, one proves the crucial fact: the intersection of any two Borel subgroups of G contains a maximal torus.

0.8. Parabolic Subgroups

A closed subgroup P of G is called **parabolic** if it contains some Borel subgroup, or equivalently, if G/P is a complete variety. Then P is connected, with $N_G(P) = P$. Each parabolic subgroup has a Levi subgroup L, a connected reductive subgroup complementary to the unipotent radical V, and all Levi subgroups of P are V-conjugate. Moreover, $P = N_G(V)$.

For a fixed Borel subgroup B, there are 2^r **standard** parabolic subgroups P_θ containing B, corresponding to subsets θ of the simple system Δ determined by B. Given a maximal torus T of B, a Levi subgroup L_θ of P_θ is then generated by T together with the root subgroups $U_{\pm\alpha}$ for $\alpha \in \theta$. So θ is a simple system for L_θ relative to T. Each parabolic subgroup of G is conjugate to precisely one of the P_θ (B14.17, H§30, S10.3).

By abuse of terminology, we refer to a parabolic subgroup as **minimal** if the semisimple rank of a Levi subgroup is 1. When a simple system Δ is fixed, the standard minimal parabolic $P_{\{\alpha\}}$ may be labelled $P_\alpha = L_\alpha V_\alpha$. If $\alpha \neq \beta$ in Δ, we have $P_\alpha \cap P_\beta = B$.

If we write $\Delta = \{\alpha_1, \dots, \alpha_r\}$, corresponding to a Borel subgroup B, the standard parabolic subgroups containing B may also be denoted $P_I = L_I V_I$, where I is a subset of $\{1, \dots, r\}$. Here the α_i $(i \in I)$ form a simple system for the Levi subgroup L_I.

0.9. Simple Algebraic Groups

A reductive group G is the product of its semisimple derived group (G,G) and a torus $S = Z(G)^\circ$, with $(G,G) \cap S$ finite. If G is a (nontrivial) semisimple group, its identity component G° is the product of uniquely determined closed connected normal subgroups G_i which are **simple** as algebraic groups: G_i is a connected semisimple group with no closed normal subgroups outside its (finite) center. (The terminology "quasi-simple" or "almost simple" is also used.) Equivalently, the root system Φ is irreducible. Here the product $G^\circ = G_1 \cdots G_n$ is "almost direct": each G_i intersects the product of the remaining ones in a finite subgroup (B14.2, B14.10, H26.2, H27.5, S9.4).

0.10. Classification

Unlike a semisimple Lie algebra over $\mathbb{C}$, a semisimple group over K is not determined by its root system Φ alone. One has to take into account the **isogeny class** of G as well, distinguishing for example between $\mathrm{SL}(n,K)$ and $\mathrm{PGL}(n,K)$. Recall that an **isogeny** $\varphi : G \to G'$ is a surjective morphism of algebraic groups, with $\operatorname{Ker}\varphi$ finite—hence central, G being connected and semisimple. (But when $\operatorname{char} K > 0$, it sometimes happens that the differential of an isogeny has a noncentral kernel in the Lie algebra $\mathfrak{g}$.)

Chevalley's fundamental classification theorem can be formalized as follows. Fix a maximal torus T in G, with character group $X = X(T)$ and cocharacter

e_α $(\alpha \in \Phi)$. Define a subalgebra of $\mathfrak{t}$ by:

$$\mathfrak{z} := \bigcap_{\alpha \in \Phi} \operatorname{Ker} d\alpha.$$

It is clear that $\mathfrak{z}$ lies in the center of $\mathfrak{g}$. (In fact, more careful analysis shows that it is precisely the center. Case-by-case study shows that $\mathfrak{z}$ has dimension at most 1 except in type D_r (r even) when $p = 2$—here the dimension is 2. See 0.13.)

All we actually need to know about $\mathfrak{z}$ is that it is fixed pointwise by all $\operatorname{Ad} x$, $x \in G$. Since G is generated by the 1-dimensional root subgroups $U_\alpha = \{x_\alpha(c) \mid c \in K\}$, it is enough to check that for $t \in \mathfrak{t}$,

$$(\operatorname{Ad} x_\alpha(c))(t) = t - c\, d\alpha(t)\, e_\alpha.$$

This follows from the standard formula for the differential of a commutator map (see B3.16 or H10.1): Namely, define $\theta_x : G \to G$ by $\theta_x(y) = yxy^{-1}x^{-1}$. Then $(d\theta_x)_e = 1 - \operatorname{Ad} x$. In particular, when $x = x_\alpha(c)$ and $t \in T$, we get

$$\theta_x(t) = tx_\alpha(c)t^{-1}x_\alpha(-c) = x_\alpha(c(\alpha(t) - 1)).$$

So $\theta_x : T \to U_\alpha$ has differential $1 - \operatorname{Ad} x$. But using the canonical identification of U_α with K, θ_x is just the map $t \mapsto c(\alpha(t) - 1)$, whose differential maps $t \in \mathfrak{t}$ to $c\, d\alpha(t)$ (or this multiple of e_α in $\mathfrak{u}_\alpha$). Comparing these two expressions for $d\theta_x$, we get the desired formula.

0.13. Ideals in $\mathfrak{g}$

In our main development, only the features of $\mathfrak{g}$ just discussed will play an essential role. But case-by-case study of the simple types leads to much more precise results, which we summarize here for the reader's convenience. Following some partial results in Humphreys [**81**], Hogeweij [**75**] and Hiss [**72**] obtained a rather complete picture of the ideal structure in $\mathfrak{g}$ and the adjoint action on $\mathfrak{g}$.

(A_r) There are distinct groups $G_{\mathrm{sc}} = \mathrm{SL}(r+1, K)$ and $G_{\mathrm{ad}} = \mathrm{PGL}(r+1, K)$, and various intermediate groups obtained by factoring out subgroups of the center of G_{sc}. If p does not divide $r+1$, then $\mathfrak{g}$ is simple and isomorphic to $(r+1, K)$. But if $p|(r+1)$, the structure of $\mathfrak{g}$ depends on the isogeny class of G. For $p = 2$ and $r = 1$, the Lie algebra of $\mathrm{SL}(2, K)$ is nilpotent: its center and derived algebra coincide, and have dimension 1. The Lie algebra of $\mathrm{PGL}(2, K)$ is solvable, but centerless. When $r > 1$ and $p|(r+1)$, $\operatorname{Lie} G_{\mathrm{sc}}$ equals its derived algebra but has a 1-dimensional center, while $\operatorname{Lie} G_{\mathrm{ad}}$ has derived algebra of codimension 1 but is centerless. In case $p^2|(r+1)$, there is an intermediate type of group whose Lie algebra is a direct sum of its 1-dimensional center and its derived algebra.

(B_r) There are distinct groups G_{sc} and G_{ad}, with isomorphic (and simple) Lie algebras unless $p = 2$. When $p = 2$, $\operatorname{Lie} G_{\mathrm{sc}}$ has a 1-dimensional center (and a derived algebra of codimension 2 when $r = 2$), while $\operatorname{Lie} G_{\mathrm{ad}}$ has derived algebra of codimension 1. There are also intermediate ideals involving root vectors for short roots, related to Chevalley's exceptional isogeny.

(C_r) There are distinct groups G_{sc} and G_{ad}, with isomorphic (and simple) Lie algebras unless $p = 2$. When $p = 2$, $\operatorname{Lie} G_{sc}$ has a 1-dimensional center and a derived algebra of dimension $2r$, while $\operatorname{Lie} G_{ad}$ has derived algebra of codimension 1. There are further intermediate ideals, as in type B.

(D_r) Besides the simply connected and adjoint groups, there are intermediate groups (two non-isomorphic types if r is even, one if r is odd). The center of $\mathfrak{g}$ is nonzero in the simply connected and intermediate cases, of dimension 2 for G_{sc} when r is even and dimension 1 otherwise. The derived algebra of $\mathfrak{g}$ is proper in the intermediate and adjoint cases, of codimension 2 for G_{ad} when r is even and codimension 1 otherwise. In odd characteristic $\mathfrak{g}$ is simple.

(E_6, E_7) In each case there are two distinct groups G_{sc} and G_{ad}. Their Lie algebras are isomorphic, and simple, with two exceptions: for type E_6 and $p = 3$, or for type E_7 and $p = 2$, $\operatorname{Lie} G_{sc}$ has a 1-dimensional center, while the derived algebra of $\operatorname{Lie} G_{ad}$ has codimension one.

(E_8, F_4, G_2) In each case $G_{sc} = G_{ad}$. Moreover, $\mathfrak{g} = [\mathfrak{g}, \mathfrak{g}]$ and $\mathfrak{z}(\mathfrak{g}) = 0$. Other ideals occur in only two cases, related to Chevalley's exceptional isogenies: type G_2 when $p = 3$ and type F_4 when $p = 2$. Here the root vectors corresponding to short roots generate an $\operatorname{Ad} G$-invariant ideal.

0.14. Morphisms

Most of the algebraic geometry needed here is fairly elementary and classical—though in later chapters we shall have to quote a few sophisticated theorems from Grothendieck's EGA. Our geometric arguments typically involve the following basic facts about the dimension of the fibres and image of a morphism:

Let $\varphi : X \to Y$ be a dominant morphism, where X and Y are irreducible varieties. Set $d := \dim X - \dim Y$. Then (a) $\varphi(X)$ contains a nonempty open subset of Y; (b) if $y \in \varphi(X)$, each irreducible component of $\varphi^{-1}(y)$ has dimension at least d, with equality holding for y in a nonempty open subset of Y; (c) if φ is bijective, then the extension of function fields $K(X)/\varphi^* K(Y)$ is finite and purely inseparable (B, AG10.1, H§4, S4.1.6).

0.15. A Key Geometric Argument

For any connected algebraic group G, the union of conjugates of a Cartan subgroup (the identity component in the centralizer of a maximal torus) contains a nonempty open subset of G, and is therefore dense in G (B11.10, H22.2, S7.3.3). When G is *reductive*, any maximal torus is equal to its centralizer (B13.17, H26.2, S9.3.5); thus the set of semisimple elements is dense in G and contains a nonempty open subset of G.

1.2. Example: General and Special Linear Groups

Before considering arbitrary semisimple groups, it is instructive to look at $\mathrm{SL}(n, K)$ together with the closely related reductive group $\mathrm{GL}(n, K)$. Since these groups differ only by the central torus in $\mathrm{GL}(n, K)$ consisting of scalar matrices, most (but not all!) questions about classes and centralizers in $\mathrm{SL}(n, K)$ can be answered by studying $\mathrm{GL}(n, K)$. Here conjugacy is the same as similarity of matrices, and we can apply the usual canonical form theory to describe conjugacy classes. In particular, since K is algebraically closed, the Jordan form of a matrix provides a convenient class representative.

In $\mathrm{GL}(n, K)$ we have the further advantage that the centralizer of an element is the principal open set consisting of invertible elements of the centralizer in $\mathrm{M}(n, K)$. Thus *all centralizers of elements in* $\mathrm{GL}(n, K)$ *are connected*, a fact which does not carry over to $\mathrm{SL}(n, K)$ (as will be illustrated in 1.12 below).

This close relationship with the full matrix algebra also makes it easy to compute the dimensions of centralizers, using just linear algebra: Find the dimension of the subalgebra of $\mathrm{M}(n, K)$ commuting with a given matrix. If the matrix lies in $\mathrm{GL}(n, K)$, its centralizer there will be a principal open subset of the commuting algebra and thus have the same dimension. Fortunately, this computation will adapt at once to $\mathrm{SL}(n, K)$: for its elements, the centralizers in the two groups obviously differ in dimension by 1.

In place of the Jordan form, we actually find it more convenient to use the rational canonical form (for which K can be arbitrary). Recall how this works for an endomorphism x of $V := K^n$. Write $c(x, V)$ and $m(x, V)$ respectively for the characteristic and minimal polynomials. Now V is the direct sum of cyclic submodules $V_1, \ldots, V_d$ relative to x, where $m(x, V_i)$ divides $m(x, V_{i+1})$. Moreover, $m(x, V_d) = m(x, V)$ and $c(x, V) = \prod m(x, V_i)$. In particular, V itself is cyclic if and only if $m(x, V) = c(x, V)$. Set $n_i := \dim V_i$, so $n_i \geq n_j$ whenever $i > j$.

An endomorphism of V commuting with x is a direct sum of homomorphisms between the various cyclic submodules V_i. Now V_j is a quotient of V_i whenever $i > j$. It follows that the dimension of $\mathrm{Hom}_x(V_i, V_j)$ is just $\min(n_i, n_j)$: When $i = j$, such a homomorphism is determined by sending a fixed generator of V_i to any vector of V_i. Similarly, when $i > j$, such homomorphisms are produced by mapping a generator of V_i to any vector in V_j. When $i < j$, one has to map V_i into the unique cyclic submodule of V_j isomorphic to V_i, so $\mathrm{Hom}_x(V_i, V_j)$ is isomorphic to $\mathrm{Hom}_x(V_i, V_i)$. Therefore the centralizer of x in $\mathrm{End}\, V$ has dimension

$$\sum_{i,j=1}^{d} \min(n_i, n_j) = \sum_{i=1}^{d} (2d - 2i + 1) n_i,$$

thanks to the simple combinatorial fact:

$$\#\{(i, j) \mid 1 \leq i, j \leq d \text{ and } \min(i, j) = k\} = 2d - 2k + 1.$$

Several interesting conclusions can be drawn. For one thing, the centralizer dimension is congruent (modulo 2) to $\sum n_i = n$, as is the dimension of $\mathrm{GL}(n, K)$. From this it follows that *the dimension of each conjugacy class in* $\mathrm{GL}(n, K)$ (*or* $\mathrm{SL}(n, K)$) *is even.* This will turn out—after considerable labor—to be true for all semisimple groups.

The formula shows also that the centralizer dimension is at least equal to the rank n (resp. $n-1$) for $x \in \mathrm{GL}(n, K)$ (resp. $\mathrm{SL}(n, K)$), with equality if and only if $d = 1$, i.e., $m(x, V) = c(x, V)$. When equality holds, the centralizer of x in $\mathrm{M}(n, K)$ is therefore just the ring of polynomials in x (whose dimension is always given by the degree of the minimal polynomial). For a semisimple matrix, the "regularity" condition $m(x, V) = c(x, V)$ means that all eigenvalues of x are distinct. For a unipotent matrix, the condition means that x is similar to a single Jordan block with 1's on and just above the main diagonal. (So the "regular" unipotent elements form a single conjugacy class.)

These elementary observations are suggestive of some of the results to be found for all semisimple groups, but the reader should be warned that both the results and the methods will become more complicated as we go along. In many respects, the special linear group is a misleadingly easy—though very important—example. It should also be emphasized that the *structure* of a centralizer as an algebraic group is not so obvious even in this special setting.

Remark. The centralizer of a single matrix (or linear operator) was studied classically by Frobenius and others. A concise modern account, in the setting of modules over principal ideal domains, may be found in the text of Adkins and Weintraub [**1**, pp. 335–337].

1.3. Centralizers of Nilpotent and Unipotent Matrices

Before leaving $\mathrm{SL}(n, K)$, we should look more closely at the dimensions of the centralizers of unipotent elements. Since x is unipotent if and only if $x - 1$ is nilpotent, and since both matrices have the same centralizer in $\mathrm{M}(n, K)$, it is enough to study the nilpotent case. Here the rational canonical form (with 1's just below the diagonal) is the transpose of the Jordan canonical form.

We used above the convention that V is the direct sum of cyclic modules of dimensions $n_1, \ldots, n_d$, where $n_1 \leq n_2 \leq \cdots \leq n_d$. Switching to the Jordan form, it is more convenient to reverse the blocks and label their sizes instead as $\lambda_1 \geq \lambda_2 \geq \cdots \geq \lambda_d > 0$. Here $\lambda := [\lambda_1, \ldots, \lambda_d]$ is a **partition** of n having d parts. With this notation, the computation in 1.2 shows that the centralizer in $\mathrm{M}(n, K)$ of our nilpotent matrix x has dimension

$$\sum_{i=1}^{d} (2d - 2i + 1)\lambda_{d-i+1} = \sum_{j=1}^{d} (2j - 1)\lambda_j.$$

This can be rewritten in a more attractive way, by using the standard identity

$$k^2 = \sum_{i=1}^{k} (2i - 1).$$

Recall that to each partition λ of n is associated its **conjugate** partition λ', where λ'_j is the number of indices i for which $\lambda_i \geq j$. For example, $\lambda'_1 = d$ here. (The terms "dual" and "transpose" are also commonly used.) This is easily visualized in terms of the **Young diagram** of λ, a left-aligned array of empty boxes with λ_1 boxes in the first row, λ_2 boxes in the second row, etc. To get the Young diagram of λ', one just transposes the diagram of λ.

It is now a simple combinatorial exercise (left to the reader) to transform the above dimension formula:

PROPOSITION. *Let $x \in \mathrm{M}(n, K)$ be a nilpotent or unipotent matrix, whose Jordan form involves blocks of sizes $\lambda_1 \geq \lambda_2 \geq \dots$ If $\mu := \lambda'$ is the conjugate partition of n, the dimension of the centralizer of x in $\mathrm{M}(n, K)$ is $\sum \mu_i^2$.*

A small numerical example should make the combinatorial argument transparent. Say $\lambda = [5, 4, 2]$ is a partition of 11, so $\mu = [3, 3, 2, 2, 1]$. The original dimension formula yields $1 \cdot 5 + 3 \cdot 4 + 5 \cdot 2 = 27$. This can be pictured using the Young diagram of λ: place 1's in the boxes in the first row, 3's in the second row, and 5's in the third row. So 27 is the sum over all boxes. If we do the sum instead by columns, we get 5 sums of the type $1 + 3 + 5$ or $1 + 3$ or just 1, corresponding to the various μ_i^2.

1.4. Orbits and Isotropy Groups

From now on G will denote an arbitrary connected semisimple group. To put the study of conjugacy classes in perspective, it is useful to recall the general situation in which an algebraic group G acts rationally on a variety X. For each $x \in X$, the isotropy group $G_x := \{g \in G \,|\, g \cdot x = x\}$ is closed in G and the orbit $G \cdot x := \{g \cdot x \,|\, g \in G\}$ is locally closed in X. The orbit map induces a bijective morphism $G/G_x \to G \cdot x$. If $G \cdot x$ is not closed, its closure is the union of $G \cdot x$ and other orbits of strictly smaller dimension. In particular, orbits of least dimension are closed.

Somewhat more subtle is an observation based on the principle of "upper semicontinuity of dimension". Recall how this principle is formulated in general (see H4.4): Given a dominant morphism of irreducible varieties $\varphi : X \to Y$, let $\varepsilon(x)$ be the largest dimension of any component of $\varphi^{-1}(\varphi(x))$ containing x. Then for all n, the set $\{x \in X \,|\, \varepsilon(x) \geq n\}$ is closed in X. (The proof adapts easily to the case when X is not required to be irreducible.) We apply this to group actions, following the arguments in Borho–Kraft [**20**, 2.1]. Here G could be any algebraic group.

PROPOSITION. *Let the algebraic group G act rationally on an irreducible variety X. Set $X_d := \{x \in X \mid \dim G_x \geq d\}$. Then X_d is closed in X for each $d \geq 0$.*

PROOF. As the inverse image of the (closed) diagonal in $X \times X$, under the morphism $(g,x) \mapsto (g \cdot x, x)$, the set

$$Z := \{(g,x) \mid g \cdot x = x\}$$

is closed in $G \times X$. Let $\pi : Z \to X$ be the restriction to Z of the projection map. Since $(e,x) \in Z$ for any $x \in X$, π is surjective. The fibre $\pi^{-1}(\pi(g,x)) = \pi^{-1}(x)$ is of the form $G_x \times \{x\}$, so its irreducible components are of equal dimension. Thanks to upper semicontinuity of dimension, the set

$$Z_d := \{(g,x) \in Z \mid \dim \pi^{-1}(x) \geq d\}$$

is closed in Z. Since π is an open map, and Z_d is a union of fibres of π, its image $\pi(Z_d) = X_d$ is therefore closed in X. □

The conclusion could equally well be stated in terms of orbit dimensions: for each d, the set $\{x \in X \mid \dim G \cdot x \leq d\}$ is closed. Equivalently, the elements of orbit dimension strictly larger than d form an open set. Call $x \in X$ **regular** if its G-orbit has the largest possible dimension. It follows that:

COROLLARY. *The set of regular elements in X is open.* □

1.5. Classes and Centralizers

We can apply the observations in 1.4 (compare B9.1) to the special case in which a connected group G acts on itself by conjugation. Write $\mathrm{Cl}(x) := \{gxg^{-1} \mid g \in G\}$, the class of x. In case H is a closed subgroup of G, it will sometimes be convenient to consider the H-class $\mathrm{Cl}_H(x) := \{gxg^{-1} \mid g \in H\}$. Write $C_G(x) := \{g \in G \mid gxg^{-1} = x\}$, the centralizer of x.

PROPOSITION. *For any $x \in G$, the conjugacy class $\mathrm{Cl}(x)$ is a smooth, irreducible, locally closed subset of G, of dimension equal to the codimension of $C_G(x)$ in G. Its closure is the union of $\mathrm{Cl}(x)$ and possibly other classes of strictly lower dimension.* □

Classes may or may not be closed in G; this issue will be clarified in 1.7 below.

A more delicate issue is the relationship between $\mathrm{Cl}(x)$ and the quotient variety $G/C_G(x)$. To insure that these are isomorphic, one needs to show that the natural morphism $G \to \mathrm{Cl}(x)$ is *separable*. While this is automatic in characteristic 0, there are exceptions in characteristic $p > 0$. As discussed in B9.1, separability turns out to be equivalent to having a good correspondence between global and infinitesimal centralizers: see 1.10 below.

The nature of the variety $G/C_G(x)$ depends of course on the nature of the group $C_G(x)$. When G is reductive, it is known that the quotient is affine if and only if $C_G(x)$ is reductive (this occurs whenever x is semisimple, as discussed in 2.2 below), whereas it is projective if and only if $C_G(x)$ is a parabolic subgroup (this very rarely occurs!).

1.6. Minimum Dimension of a Centralizer

An element $x \in G$ is **regular** when the dimension of $\mathrm{Cl}(x)$ is as large as possible, or equivalently, when $\dim C_G(x)$ is as small as possible. As shown in 1.4, the regular elements of G form an open subset, which we denote by G_{reg}. The rank of G affords a natural lower bound on the dimension of all centralizers:

PROPOSITION. *For all* $x \in G$, $\dim C_G(x) \geq r$, *where* r *is the rank of* G. *More precisely,* $\dim C_B(x) \geq r$ *for any Borel subgroup* B *containing* x.

PROOF. Let B be any Borel subgroup containing x, with unipotent radical U. From 0.3 we know that $(B, B) \subset U$ (in fact, equality holds). Therefore $(B, x) \subset U$, and $\mathrm{Cl}_B(x) \subset Ux$, so its dimension is at most $\dim U$. In turn, $\dim C_B(x) \geq \mathrm{codim}_B\, U = r$, forcing $\dim C_G(x) \geq r$. □

In later chapters we shall study regular elements in detail, showing in particular that their centralizers are of dimension exactly r (see 2.3). It will be easy to verify the existence of regular semisimple elements, but considerably more difficult to do so in the case of unipotent elements.

1.7. Closures of Conjugacy Classes

It is a fundamental fact that in any algebraic group, *all conjugacy classes of semisimple elements are closed* (see H18.2). (The key tangent space calculation involved there will be recalled in 3.8 below.) In a semisimple—or, more generally, reductive—group, the semisimple classes turn out to be the only closed ones:

PROPOSITION. *For any* $x \in G$, $x_s \in \overline{\mathrm{Cl}(x)}$. *In particular,* e *lies in the closure of every unipotent class.*

PROOF. Set $C := \mathrm{Cl}(x)$. We may assume that x belongs to a Borel subgroup $B = TU$, with $x_s \in T$, $x_u \in U$. Write

$$x_u = \prod_{\alpha>0} x_\alpha(c_\alpha)$$

for some $c_\alpha \in K$. If $c \in K^\times$ is arbitrary, we can find $t \in T$ for which $\alpha(t) = c$ for all *simple* roots α. (See H16.2, Lemma C.) In turn, $txt^{-1} = x_s(tx_ut^{-1}) = x_s u_c$, where

$$u_c := \prod_{\alpha>0} x_\alpha(c^{\mathrm{ht}\,\alpha} c_\alpha).$$

Here $\mathrm{ht}\,\alpha$ is the sum of coefficients of α when written as a $\mathbb{Z}$-linear combination of simple roots. Note that u_c makes sense even when $c = 0$: then $u_0 = e$.

Now we have $u_c \in x_s^{-1}C$ for all nonzero c. For any $f \in K[G]$ vanishing on $x_s^{-1}C$, the polynomial map sending c to $f(u_c)$ vanishes on $K^\times$, hence vanishes at $c = 0$. In particular, $f(e) = f(u_0) = 0$, forcing $e \in \overline{x_s^{-1}C} = x_s^{-1}\overline{C}$, or $x_s \in \overline{C}$ as required. □

COROLLARY. Cl(x) *is closed (if and) only if* x *is semisimple.* □

For analogous results on semisimple elements of $\mathfrak{g}$, see B9.2 or S4.4.5.

Remark. Ferrer Santos [**53**] looks more closely at the algebraic groups H for which all classes are closed, and proves that—at least in characteristic 0—this condition forces H° to be nilpotent. The idea is to find a Levi subgroup, a reductive complement to the unipotent radical of H (which need not exist in prime characteristic). Then observe that each class of H intersects the Levi subgroup in a single class, which is therefore closed by the assumption on H, hence must be a semisimple class by the above corollary. This forces the Levi subgroup to be a torus, and H° to be solvable. Further study of the solvable case (valid in arbitrary characteristic) leads to the conclusion that H° is nilpotent. More generally, Ferrer Santos studies groups whose orbits are all closed in any action on an affine variety.

1.8. Morphisms and Subgroups

The following proposition, whose proof is immediate, sometimes permits us to replace G by a closely related group (such as a group in the same isogeny class) without loss of essential information: for example, when we investigate the finiteness of the number of unipotent classes.

PROPOSITION. *Let* $\varphi : G \to G'$ *be an epimorphism of algebraic groups, with kernel* N.

(i) *The pre-image in* G *of any class of* G' *is a union of classes (finitely many if* N *is finite).*
(ii) *If* N *is central (and thus consists of semisimple elements),* φ *induces a bijection between the unipotent classes of* G *and the unipotent classes of* G'. □

There are also some easy observations to be made about the relationship between classes of G and classes of a closed subgroup H. For example, if $G = G_1 \times \cdots \times G_n$ is a direct product, and $x = x_1x_2\cdots x_n$ with $x_i \in G_i$, then $C_G(x) = C_{G_1}(x_1) \times \cdots \times C_{G_n}(x_n)$. This allows one to reduce many questions about semisimple groups to the simple case.

When H is a closed subgroup of G, each class of G intersects H in a union of classes of H. In this case, it is usually interesting to ask whether the number of classes is finite or not. See for example 3.8 below.

CHAPTER 2

Centralizers of Semisimple Elements

In this chapter we explore semisimple conjugacy classes, focusing mainly on the structure of the associated centralizers. Unless otherwise stated, G always denotes a connected semisimple group of rank r.

First we lay out the structure of a centralizer (2.2). This leads to a better picture of regular semisimple elements: for example, they form a dense open subset of G (2.5). In case G is *simply connected*, we show that all centralizers of semisimple elements are in fact connected (2.11). Finally, we discuss some characterizations of centralizers among the reductive subgroups of maximal rank.

2.1. Subgroups of Maximal Rank

If $s \in G$ is semisimple, it lies in some maximal torus T, and thus $T \subset C_G(s)$. So the centralizers of semisimple elements are among the subgroups of maximal rank: subgroups containing a maximal torus. As we shall see in the following section, $C_G(s)$ is also reductive—though not necessarily connected.

Other natural examples of reductive subgroups of maximal rank are the Levi factors of parabolic subgroups (H§30) and the centralizers of arbitrary tori (H26.2). These are always connected. In fact, a closer scrutiny of closed subsets of the root system shows that centralizers of tori are precisely the Levi factors of parabolic subgroups: see [**17**, 4.15] or [**46**, 1.22] for details. While our main concern is with centralizers of semisimple elements, it may be helpful to glance briefly at the general picture.

Using some of the standard structure theory of reductive groups, it is not difficult to describe all connected subgroups of maximal rank (see B13.20 for details): If H contains the maximal torus T, the complete reducibility of $\operatorname{Ad} T$ acting on $\mathfrak{g}$ implies that $\mathfrak{h}$ is the direct sum of $\mathfrak{t}$ and various one-dimensional root spaces. So the root system Ψ of H relative to T consists of a subset of the root system of G relative to T. (For example, there is a subgroup of type A_2 in a group of type G_2 corresponding to the set of long roots.) Now one can argue that H itself is generated by T along with the root groups U_α ($\alpha \in \Psi$), using the especially good correspondence between global and infinitesimal centralizers of diagonalizable groups such as the kernels of roots.

Each $c \in K^\times$ lies in some finite subfield of order p^e, whose multiplicative group is cyclic of order $p^e - 1$; thus c is a root of unity of order relatively prime to p. More precisely, for a given integer $s > 0$ prime to p, the group C_s of all sth roots of unity in $K^\times$ is cyclic of order s: the polynomial $x^s - 1$ has s distinct roots in K (by the usual derivative test), all lying in some finite subfield, whose multiplicative group is cyclic. In turn, $K^\times$ is the union of the subgroups C_s ($p \not| s$). In fact we can choose a sequence $s_1, s_2, \ldots$ for which $C_{s_1} \subset C_{s_2} \subset \cdots$ and $K^\times = \bigcup C_{s_i}$. To be specific, take s_i to be the p'-part of $i!$.

We invoke an elementary fact about cyclic groups: If C is cyclic of order $n = st$, and D is the subgroup generated by an element d of order t, then $d = c^s$ for some generator c of C. To prove this, start with any generator a of C, write $d = a^k$, and let $m = \gcd(k, n)$. If $k = lm$, it follows from $\gcd(l, n) = 1$ that $c = a^l$ is another generator of C, and $c^m = a^k = d$. But since $m|n$, c^m has order n/m, forcing $n/m = t$ or $m = s$. Thus $d = c^s$.

Using this fact, we can find generators c_i of C_{s_i} for all i so that c_i is the s_{i+1}/s_i power of c_{i+1}. Now we can define a group homomorphism $\varphi : \mathbb{Q}_{p'} \to K^\times$. Given a/b with $a, b \in \mathbb{Z}$ and $p \not| b$, find s_i divisible by b and rewrite $a/b = m/s_i$. Let $\varphi(a/b) = c_i^m$. This is independent of the choice of s_i divisible by b, thanks to the compatibility of the generators c_i. By choosing a common denominator for two fractions, one checks that φ is a group homomorphism. It is surjective, since $\varphi(1/s_i) = c_i$, and its kernel is clearly $\mathbb{Z}$.

To summarize:

LEMMA. *Let K be an algebraically closed field.*

(i) *If $K = \mathbb{C}$, the group Ω of roots of unity in $K^\times$ is isomorphic to the additive group $\mathbb{Q}/\mathbb{Z}$.*

(ii) *If K is an algebraic closure of $\mathbb{F}_p$ for a prime p, the group $K^\times$ of roots of unity in K is isomorphic to the additive group $\mathbb{Q}_{p'}/\mathbb{Z} = (\mathbb{Q}/\mathbb{Z})_{p'}$.* □

2.8. Tori and Characters

PROPOSITION. *Let T be a maximal torus of G, and let $Y = Y(T)$ be its group of cocharacters. There is an isomorphism of abelian groups $\theta : Y \otimes K^\times \to T$ sending $\lambda \otimes c \mapsto \lambda(c)$.*

PROOF. It is clear that $(\lambda, c) \mapsto \lambda(c)$ is a balanced map which yields the group homomorphism θ. In the other direction, choose a basis $\lambda_1, \ldots, \lambda_r$ of Y. Each $t \in T$ can then be written uniquely as $\prod \lambda_i(c_i)$ with $c_i \in K^\times$. So we just send $t \mapsto \sum \lambda_i \otimes c_i$ to get an inverse for θ. □

There is a more detailed treatment in Carter [**29**, 3.1], showing that $Y \otimes K^\times$ is isomorphic to $\mathrm{Hom}(X, K^\times)$, where $X = X(T)$, and in turn that $\mathrm{Hom}(X, K^\times)$ is isomorphic to T.

In case K is an algebraic closure of $\mathbb{F}_p$, we can combine this identification with Lemma 2.7 to obtain an isomorphism $T \cong Y \otimes (\mathbb{Q}/\mathbb{Z})_{p'}$. By using a common denominator for the rational numbers occurring in a sum of tensors, we can then identify an element $s \in T$ with a single tensor $\lambda \otimes q$, where $q \in (\mathbb{Q}/\mathbb{Z})_{p'}$. The numerator of a fraction can be combined with λ, allowing us to represent q by a fraction $1/d$ with d an integer not divisible by p. For brevity, we may even write λ/d to represent s. Concretely, s is an element of finite order d (not divisible by p), obtained by applying $\lambda : K^\times \to T$ to a primitive dth root of unity in $K^\times$ which corresponds to the residue class of $1/d$ under our chosen isomorphism between $K^\times$ and the additive group $(\mathbb{Q}/\mathbb{Z})_{p'}$.

In a similar spirit, when $K = \mathbb{C}$, and s is an element of finite order d in T, we can write $s = \lambda/d$ for some $\lambda \in Y$.

2.9. The Dual Group of a Reductive Group

Here G may be any connected reductive group. There is a naturally defined "dual" group, also connected and reductive, whose root system is the coroot system of G relative to a maximal torus. This type of duality is important in the Langlands program, and when combined with the study of Frobenius endomorphisms in prime characteristic, is critical to the Deligne–Lusztig character theory for finite groups of Lie type (see Carter [**29**, Chap. 4]).

The dual group arises as follows. Fix a maximal torus T of G, together with its character group $X = X(T) = \mathrm{Hom}(T, K^\times)$, a free abelian group of rank $r = \mathrm{rank}\, G$. Let $\Phi \subset X$ be the set of roots of G relative to T. Similarly, write $Y = Y(T) = \mathrm{Hom}(K^\times, T)$, the group of cocharacters, and let $\Phi^\vee \subset Y$ be the set of coroots. Now the quadruple $(X, \Phi, Y, \Phi^\vee)$ is the root datum of G relative to T (0.10). By the fundamental theorem of Chevalley, it determines G uniquely, up to isomorphism.

The quadruple $(Y, \Phi^\vee, X, \Phi)$ obtained by interchanging the roles of X and Y is another root datum. By the fundamental existence theorem, it determines another connected reductive group. Call this group G^* the **dual group** of G. It is uniquely determined, up to isomorphism, and has a maximal torus T^* with character group Y and cocharacter group X.

For example, the dual group of $\mathrm{GL}(n, K)$ turns out to be $\mathrm{GL}(n, K)$ itself. In the case of simple groups, duality just interchanges simply connected and adjoint groups of the same type, with one interesting exception: the dual of the simply connected group $\mathrm{Sp}(2r, K)$ of type C_r is the adjoint group $\mathrm{SO}(2r+1, K)$ of type B_r, while the dual of the adjoint group $\mathrm{PSp}(2r, K)$ is the simply connected group $\mathrm{Spin}(2r+1, K)$. The groups of types E_8, F_4, G_2 are self-dual, being simultaneously simply connected and adjoint groups. (For lists of dual groups over finite fields see Carter [**28**, p. 3] or [**29**, p. 120].)

2.10. Semisimple Elements of Finite Order

We shall also need the result of a standard exercise in Bourbaki [**23**, VI, §2, exer. 1], which is worked out in Jantzen [**87**, 1.3]. This involves the action of the Weyl group W of G on the real vector space $V := X(T) \otimes \mathbb{R}$, and can be formulated just in the context of reflection groups and abstract root systems.

LEMMA. *For any $\lambda \in V$, the subgroup $W_\lambda := \{w \in W \mid w\lambda - \lambda \in L(\Phi)\}$ of W is generated by the reflections s_α $(\alpha \in \Phi)$ which it contains.*

PROOF. The trick is to introduce the affine Weyl group W_a, the semidirect product of W and the lattice of translations in V corresponding to the root lattice $L(\Phi)$. The group W_a is generated by affine reflections. Moreover, the isotropy group of λ in W_a is generated by the affine reflections it contains (see Bourbaki [**23**, V, §3, Prop. 1]). These affine reflections are given by

$$s_{\alpha,n}(\lambda) = s_\alpha\lambda + n\alpha = \lambda + (n - \langle\lambda, \alpha^\vee\rangle)\alpha,$$

for $\alpha \in \Phi$ and $n \in \mathbb{Z}$. We conclude that $s_{\alpha,n}(\lambda) = \lambda$ for some $n \in \mathbb{Z}$ if and only if $\langle\lambda, \alpha^\vee\rangle \in \mathbb{Z}$, or if and only if $s_\alpha\lambda - \lambda \in L(\Phi)$. Thus W_λ is generated by those reflections s_α for which some corresponding affine reflection $s_{\alpha,n}$ fixes λ. □

Following Digne–Michel [**46**, 13.14], we now look closely at the quotient $W(s)/W^\circ(s)$ for a semisimple element s of G whose order is finite.

PROPOSITION. *Let $s \in G$ be a semisimple element of finite order, lying in the maximal torus T with Weyl group W. If T^* is the corresponding maximal torus of the dual group G^*, the quotient $W(s)/W^\circ(s)$ is isomorphic to a subgroup of $X(T^*)/L(\Phi^\vee)$, of order prime to p if* $\operatorname{char} K = p > 0$. *In particular, if G is simply connected, $W(s) = W^\circ(s)$ and $C_G(s)$ is connected.*

PROOF. (i) Suppose $\operatorname{char} K = p > 0$ and s has finite order d in G. Let s lie in the maximal torus T, and let T^* be the maximal torus of the dual group G^* having character group $Y = Y(T)$. Using the identifications worked out in 2.8, we have $T \cong Y \otimes K^\times \cong X(T^*) \otimes K^\times$. Identify s with χ/d, where $\chi \in X(T^*)$ and $\gcd(p, d) = 1$. Then we can characterize

$$W(s) = \{w \in W \mid w\chi - \chi \in dX(T^*)\}.$$

The key step is to define a group homomorphism $\psi : W(s) \to X(T^*)/L(\Phi^\vee)$ by the recipe

$$w \mapsto \frac{w\chi - \chi}{d} \mod L(\Phi^\vee).$$

This really is a homomorphism:

$$\frac{ww'\chi - \chi}{d} = \frac{ww'\chi - w\chi}{d} + \frac{w\chi - \chi}{d} \equiv \frac{w'\chi - \chi}{d} + \frac{w\chi - \chi}{d} \mod L(\Phi^\vee),$$

using the fact that W acts trivially on $X(T^*)/L(\Phi^\vee)$ to get

$$\frac{ww'\chi - w\chi}{d} = w\left(\frac{w'\chi - \chi}{d}\right) \equiv \frac{w'\chi - \chi}{d} \mod L(\Phi^\vee).$$

We have $\operatorname{Ker}\psi = \{w \in W \mid w(\chi/d) - (\chi/d) \in L(\Phi^\vee)\}$. The above lemma can be applied to the root system $\Phi^\vee$ in $X(T^*) \otimes \mathbb{R}$, whose Weyl group is also W, to conclude that $\operatorname{Ker}\psi$ is generated by various reflections $s_\alpha = s_{\alpha^\vee}$.

Now $s_\alpha\chi - \chi = \langle \chi, \alpha \rangle \alpha^\vee$, so $s_\alpha \in \operatorname{Ker}\psi$ if and only if $\langle \chi, \alpha \rangle \in d\mathbb{Z}$, or if and only if $\langle \chi/d, \alpha \rangle \in \mathbb{Z}$. This is equivalent to saying that $\alpha(s) = 1$, or that $s_\alpha \in W^\circ(s)$. Thus $\operatorname{Ker}\psi = W^\circ(s)$. Finally, we conclude that $W(s)/W^\circ(s)$ is isomorphic to a p'-subgroup of $X(T^*)/L(\Phi^\vee)$.

(ii) If $\operatorname{char} K = 0$, while s has finite order, the preceding argument needs only slight modifications: use the last line of 2.8 to write $s = \chi/d$ (and suppress mention of p). □

When $\operatorname{char} K = p > 0$, we have assumed that K is an algebraic closure of its prime field. But the conclusion of the proposition remains valid for any larger algebraically closed field, say L. The point is that any semisimple element of *finite order* in $G(L)$ has a conjugate in $G(K)$, since $K^\times$ contains all roots of unity in L. Moreover, Weyl group representatives can be chosen to lie in $G(K)$. So the proof above will still yield the structure of $W(s)/W^\circ(s)$.

2.11. Connectedness Theorem

Finally we can prove the main theorem. When $\operatorname{char} K = 0$, we have assumed for concreteness that $K = \mathbb{C}$; but the preceding arguments carry over readily to any algebraically closed field of characteristic 0. When $\operatorname{char} K = p$, we have assumed that K is an algebraic closure of $\mathbb{F}_p$. This too is inessential, as remarked at the end of the preceding section.

THEOREM. *If G is simply connected, then the centralizer of any semisimple element of G is connected.*

PROOF. Let s be a semisimple element. Proposition 2.10 disposes of the case in which s has finite order. If the order is infinite, consider the closure S of the cyclic subgroup generated by s. Evidently $C_G(s) = C_G(S)$. Now S is a diagonalizable group, and can therefore be written as the direct product of its identity component and a finite subgroup D (see H16.2, S2.5.8). If S is connected (hence a torus), then $C_G(S)$ is connected (see H22.3) and we are done. Otherwise, write $s = s_0 t$ for some $s_0 \in S^\circ, t \in D$. Consider the canonical map $S \to S/S^\circ \cong D$. The inverse image of the subgroup of D generated by t is closed and contains s, hence coincides with S. This shows that t generates D. Because t has finite order, we know from 2.10 that $H := C_G(t)$ is connected. But now $C_G(s) = C_G(S) = C_H(S^\circ)$. Since the centralizer of a torus in a connected group such as H is again connected, we conclude that $C_G(s)$ is connected. □

Remark. It is worth observing that our analysis of semisimple elements of finite order yields more generally: *Let G be a connected reductive group, with dual group G^*. If the center of G^* is connected, then the centralizer of any semisimple element of finite order in G is connected.*

(but not others) in $\mathfrak{g}$ can be separated by G-invariant polynomials. All of this generalizes—with some complications in small prime characteristics—to the Lie algebra of a semisimple group over any K: see Springer–Steinberg [**180**, II, 3.17′] and Slodowy [**156**, 3.12].

3.5. The Steinberg Map

In case G is *simply connected*, the adjoint quotient $\chi : G \to T/W$ takes on a more concrete appearance. Here we can identify T/W with $\mathbb{A}^r$ using the characters χ_i of the fundamental irreducible representations. (All of this is relative to a fixed choice of maximal torus and simple system of roots.) The map χ becomes the **Steinberg map** $\pi : G \to \mathbb{A}^r$, defined by $\pi(x) := (\chi_1(x), \dots, \chi_r(x))$. It is surjective and its fibres are unions of conjugacy classes, with each fibre containing a unique semisimple class.

As we observed in 3.2, when $G = \mathrm{SL}(n, K)$ the map π just associates to a matrix the significant coefficients (with alternating sign) of its characteristic polynomial. So each fibre of π consists of matrices sharing that polynomial. When we study regular elements in Chapter 4, we will be able to describe the fibres of the Steinberg map in more detail for arbitrary G. But we can already say something in this direction:

PROPOSITION. *Let G be simply connected, with Steinberg map $\pi : G \to \mathbb{A}^r$. Let F be a fibre of π, and let s be an element of the unique semisimple class contained in F. Then x lies in F if and only if x_s is conjugate to s. In particular, the fibre containing e is the set of all unipotent elements of G.*

PROOF. Let $x \in G$. Since x_s lies in the closure of $\mathrm{Cl}(x)$ (Proposition 1.7), all class functions agree at x and x_s. So x and x_s both lie in the same fibre of π. This fibre is F if and only if x_s is conjugate to s. $\square$

The last assertion of the proposition may be reformulated (in a way which remains valid even if G is not simply connected):

COROLLARY. *An element $x \in G$ is unipotent if and only if $\chi(x) = \chi(e)$ for all irreducible characters χ of G.* $\square$

3.6. Rational Class Functions

When G is *simply connected*, the Steinberg map $\pi : G \to \mathbb{A}^r$ may be regarded as a kind of "quotient" map. We shall explain this briefly, referring the reader to [**184**, §6] for details. One needs to examine not just the algebra of regular functions $K[G]$ but also the field $K(G)$ of rational functions. Here we give each rational function on G its largest possible domain of definition (an open set).

Define $C(G)$ to be the subalgebra of $K(G)$ consisting of rational functions which are constant on classes where they are defined: **rational class functions**. One has to check that a rational class function defined at one point of a class is defined everywhere on that class: compare the function obtained from the given one by applying the conjugation action of an element of G.

Obviously $C(G)$ contains all quotients of regular class functions. The reverse inclusion also holds: Since the semisimple elements are dense in G and each has a conjugate in T, the domain of definition of a given function in $C(G)$ includes a dense open subset of T. Imitating the proof of Theorem 3.2, we see that the restriction map induces an isomorphism from $C(G)$ onto the algebra $C(T/W)$ of W-invariant rational functions on T. This reduces our problem to the case of the finite group W. Let $f \in C(T/W)$ be represented by a quotient g/h of regular functions on T. We can rewrite f by g_0/h_0, where $h_0 := \prod_{w \in W} w \cdot h$ lies in $C[T/W]$ and $g_0 := g \prod_{w \neq e} w \cdot h$. Since f is W-invariant, it follows that g must be in $C[T/W]$ as well.

In this setting we can state more precisely what it means for the Steinberg map to be a quotient:

THEOREM. *Let G be simply connected, with Steinberg map $\pi : G \to \mathbb{A}^r$.*

(i) *Given a rational function f on $\mathbb{A}^r$ and an element $x \in G$, f is defined at $\pi(x)$ precisely when $f \circ \pi$ is defined at x. As a result, the domain of definition of a rational class function on G consists of complete fibres.*
(ii) *The comorphism π^* maps the function field of $\mathbb{A}^r$ isomorphically onto the field of rational functions on G constant on fibres of π.*

From this it follows that the collection of fibres of the Steinberg map has a natural structure of affine variety (isomorphic to $\mathbb{A}^r$). Restriction of π to the open subset G_{reg} then gives the same structure of affine variety to the collection of regular classes. In case G is not simply connected, Steinberg points out [**184**, 6.18] that the collection of regular classes can instead be given the structure of the (affine) quotient variety $\mathbb{A}^r/F$, where F is the "fundamental group" of G: the kernel of the natural map from $G_{\text{sc}} \to G$.

3.7. Similarity and Conjugacy

Although the characters of representations contain too little information to distinguish between classes, the representations themselves might still do the job. Obviously conjugate elements of G are represented by similar matrices (or linear operators). In his address at the 1966 International Congress [**185**, (4)], Steinberg posed the problem of proving a converse:

Elements x and y of G are conjugate in G if $\rho(x)$ and $\rho(y)$ are similar for every irreducible rational representation ρ of G.

At this writing, the problem has not been completely solved—unless both of the elements in question are semisimple (3.4) or regular (4.14 below). The analogous conjecture for a semisimple Lie algebra in characteristic 0 was proved by Gauger [**58**], then proved in another way by Steinberg [**192**]. Steinberg's method, using the Jacobson–Morozov theorem on embedding nilpotent elements into three dimensional simple subalgebras, carries over to the algebraic group provided the characteristic is 0 or else large enough relative to the Coxeter number. These results encourage the hope that the problem will always have an affirmative solution.

When $\operatorname{char} K = p > 0$, Steinberg's tensor product theorem (see Jantzen [**89**, II3.17]) implies at once that similarity of $\rho(x)$ and $\rho(y)$ for all irreducible representations ρ of **restricted** highest weight would imply similarity for all irreducible representations. The restricted dominant weights are those whose coordinates relative to fundamental weights lie between 0 and $p-1$. So there are only p^r of them ($r = \operatorname{rank} G$). Thus an affirmative solution to Steinberg's problem would immediately imply that the number of unipotent classes of G is finite. This finiteness theorem has so far been proved uniformly in characteristic p only in a very sophisticated and indirect manner (see 3.11 below). In the following sections we shall present a much more direct approach to the theorem, which unfortunately cannot be applied in some cases when $p = 2, 3, 5$.

3.8. Richardson's Finiteness Theorem

For $\mathrm{GL}(n, K)$ and $\mathrm{SL}(n, K)$, the Jordan canonical form provides convenient representatives of conjugacy classes. In either case, the unipotent classes are in natural bijection with the partitions λ of n (as in 1.3): if $n = \lambda_1 + \cdots + \lambda_d$ with $\lambda_1 \geq \cdots \geq \lambda_d > 0$, construct a corresponding unipotent matrix with Jordan blocks of dimensions $\lambda_1, \ldots, \lambda_d$. In particular, there are only finitely many unipotent classes. The following theorem, due to Richardson [**136**], provides a simple general criterion for the finiteness of the number of unipotent classes in a connected algebraic group. We shall see in the next section that it can be applied readily to "most" (but not all) semisimple groups. Here we temporarily use the letter G to denote $\mathrm{GL}(n, K)$.

THEOREM. *Let H be a closed connected subgroup of $G := \mathrm{GL}(n, K)$, whose Lie algebra $\mathfrak{h}$ satisfies the following condition:*
(*) *There exists a subspace $\mathfrak{m}$ of $\mathfrak{g}$, stable under $\operatorname{Ad} H$, for which $\mathfrak{g} = \mathfrak{h} \oplus \mathfrak{m}$. Then any conjugacy class of G intersects H in only finitely many conjugacy classes of H. In particular, H has only finitely many unipotent classes.*

PROOF. Fix a conjugacy class C_0 of G. The intersection $C_0 \cap H$ is a union of classes, each of which (being irreducible) lies in an irreducible component of the intersection. Since the number of components is finite, it will be enough to show that each component Z is itself a class of H. If C is any class of H contained in Z, it will in turn be enough to show that $\dim C = \dim Z$: then C must contain a dense open subset of Z, and (C being arbitrary) each class contained in Z must meet C, hence equal C.

The dimension comparison is carried out on the level of tangent spaces, using the special assumption (*). Fix an element $h \in C$ and define $\varphi : G \to C_0 h^{-1}$

by $\varphi(x) = xhx^{-1}h^{-1}$, so $\varphi(e) = e$. An elementary calculation shows that the differential is given by $d\varphi_e = 1 - \operatorname{Ad} h$ (see H10.1).

We claim that $d\varphi_e$ is surjective. For this, recall (1.10) that the Lie algebra of $C_G(h)$ is $\mathfrak{c}_{\mathfrak{g}}(h) = \{x \in \mathfrak{g} \mid (\operatorname{Ad} h)(x) = x\}$: the dimensions agree here because $C_G(h)$ is a principal open subset of $\mathfrak{c}_{\mathfrak{g}}(h)$. Now we compute:

$$\begin{aligned}
\dim d\varphi_e(\mathfrak{g}) &= \dim \mathfrak{g} - \dim \operatorname{Ker} d\varphi_e \\
&= \dim \mathfrak{g} - \dim \mathfrak{c}_{\mathfrak{g}}(h) \\
&= \dim G - \dim C_G(h) \\
&= \dim C_0 \\
&= \dim C_0 h^{-1} \\
&= \dim \mathrm{T}(C_0 h^{-1})_e \text{ (since } C_0 \text{ is smooth).}
\end{aligned}$$

Now we can complete the proof:

$$\begin{aligned}
\mathrm{T}(Ch^{-1})_e &\subset \mathrm{T}(Zh^{-1})_e \\
&\subset \mathrm{T}(C_0 h^{-1})_e \cap \mathrm{T}(H)_e \\
&= (1 - \operatorname{Ad} h)(\mathfrak{g}) \cap \mathfrak{h} \\
&= (1 - \operatorname{Ad} h)(\mathfrak{h}) \\
&\subset \mathrm{T}(Ch^{-1})_e,
\end{aligned}$$

using (*) to obtain the non-obvious inclusion in the next-to-last line. Thus equality must hold everywhere in this chain; in particular $\mathrm{T}(Ch^{-1})_e = \mathrm{T}(Zh^{-1})_e$. This forces $\mathrm{T}(C)_h = \mathrm{T}(Z)_h$. Since C is smooth,

$$\dim C = \dim \mathrm{T}(C)_h = \dim \mathrm{T}(Z)_h \geq \dim Z,$$

yielding $\dim C \geq \dim Z$, and finally $\dim C = \dim Z$ as desired. □

The chain of equalities near the end of the proof yields an extra dividend: the global and infinitesimal centralizers of any element of H have the same dimension (just as in the case of $\mathrm{GL}(n, K)$). To see this, note that the argument applies to any element $h \in H$ and its class C, showing that $\dim C = \dim \mathrm{T}(C)_h = \dim \mathrm{T}(Ch^{-1})_e$ is the same as the dimension of the image of $1 - \operatorname{Ad} h : \mathfrak{h} \to \mathfrak{h}$. So $\dim C_H(h) = \dim H - \dim C$ is the same as $\dim \mathfrak{h} - \dim(1 - \operatorname{Ad} h)(\mathfrak{h}) = \dim \operatorname{Ker}(1 - \operatorname{Ad} h) = \dim \mathfrak{c}_{\mathfrak{h}}(h)$.

COROLLARY. *Under the assumptions of the theorem,* $\operatorname{Lie} C_H(h) = \mathfrak{c}_{\mathfrak{h}}(h)$ *for all* $h \in H$. □

3.9. Unipotent Classes in Good Characteristic

When $\operatorname{char} K = 0$ and H is reductive, there is no difficulty about satisfying the condition (*) in Theorem 3.8: the complete reducibility of all finite-dimensional rational representations of H (see 0.11) allows one to use any embedding of H into a general linear group. But complete reducibility fails badly in prime characteristic, unless H is a torus. So a more precise choice of embedding is required. Such a choice can easily be made if we are willing to avoid the primes $2, 3, 5$ in certain cases.

Let G again denote a semisimple group. The essential situation is that in which G is *simple*, i.e., has an irreducible root system of one of the types $\mathrm{A}_r, \mathrm{B}_r, \mathrm{C}_r, \mathrm{D}_r, \mathrm{E}_6, \mathrm{E}_7, \mathrm{E}_8, \mathrm{F}_4, \mathrm{G}_2$. We say that a prime p is **bad** for G if $p = 2$ and G is of type B, C, D, E, F, G, or if $p = 3$ and G is of type E, F, G, or if $p = 5$ and G is of type E_8. Otherwise p is a **good prime** for G. (Being a bad prime is in fact equivalent to occurring as a coefficient of the highest root: see Springer–Steinberg [**180**, I, 4.3].) In turn, we say that $\operatorname{char} K$ is **good** for G if it is either 0 or else a good prime. When G is semisimple, we may say that $\operatorname{char} K$ is good for G if it is good for each simple factor.

THEOREM. *If G is simple and* $\operatorname{char} K$ *is good for G, then G has only finitely many unipotent classes.*

PROOF. We may assume that $\operatorname{char} K$ is a good prime p. It is enough to consider any group in the isogeny class of G (Proposition 1.8).

We have already settled the case of type A, when G is isogenous to $\mathrm{SL}(n, K)$. For the other classical groups, of types B, C, D (when $\operatorname{char} K \neq 2$), we need only consider the special orthogonal and symplectic groups, in their natural linear realizations, and argue that condition (*) in Theorem 3.8 applies to G (in the role of H there). In each case we have a nondegenerate bilinear form (say β) on a vector space V, and the classical group G may be taken to be the identity component of the subgroup of $\mathrm{GL}(V)$ leaving the form invariant. For any endomorphism t of V, we define its adjoint t^* as usual by the condition

$$\beta(t(u), v) = \beta(u, t^*(v)).$$

Then we can describe the Lie algebra of G as the space of skew-symmetric endomorphisms of V:

$$\mathfrak{g} = \{t \in \mathrm{End}(V) \mid t + t^* = 0\}.$$

Now set

$$\mathfrak{m} := \{t \in \mathrm{End}(V) \mid t - t^* = 0\}.$$

Clearly $\mathfrak{m}$ is stable under conjugation by G in $\mathrm{End}(V)$. Moreover, an elementary calculation shows that $\mathfrak{gl}(V) = \mathfrak{g} \oplus \mathfrak{m}$ (vector space direct sum). So condition (*) in Theorem 3.8 is met for the classical groups in characteristics different from 2.

For the exceptional types E, F, G, we adopt a different strategy. Let $H = \operatorname{Ad} G$, a subgroup of $\mathrm{GL}(\mathfrak{g})$, so $\mathfrak{h} = \operatorname{ad} \mathfrak{g} = \operatorname{Der} \mathfrak{g} \subset \mathfrak{gl}(\mathfrak{g})$. The bilinear form $\langle X, Y \rangle := \mathrm{Tr}(XY)$ on $\mathrm{End}(\mathfrak{g})$ is well-known to be nondegenerate, and its restriction to $\mathfrak{h} \cong \mathfrak{g}$ is just the Killing form. We claim that this Killing form is nondegenerate (as in characteristic 0). Seligman [**142**] shows how to calculate the determinant of the Killing form, relative to a Chevalley basis over $\mathbb{Z}$. (A similar calculation is carried out by Springer–Steinberg [**180**, p. 180].) For the exceptional types, when p is good, the determinant is not divisible by p, as shown in the table.

Type	Det
G_2	$2^{42}3^7$
F_4	$2^{78}3^{104}$
E_6	$2^{234}3^{79}$
E_7	$-2^{267}3^{266}$
E_8	$2^{496}3^{248}5^{248}$

TABLE. Determinant of Killing form

Thus the restriction to $\operatorname{ad}\mathfrak{g}$ of the nondegenerate form $\operatorname{Tr}(XY)$ on $\mathfrak{gl}(\mathfrak{g})$ is still nondegenerate, and the orthogonal complement of $\mathfrak{h} = \operatorname{ad}\mathfrak{g}$ in $\mathfrak{gl}(\mathfrak{g})$ is the desired subspace $\mathfrak{m}$ meeting Richardson's condition (*). □

3.10. Nilpotent Orbits

Richardson's Theorem 3.8 was also formulated by him in the Lie algebra setting:

THEOREM. *Let G and H be as in Theorem* 3.8, *satisfying condition* $(*)$ *there. Then any G-orbit in $\mathfrak{g}$ intersects $\mathfrak{h}$ in only finitely many H-orbits. In particular, $\mathfrak{h}$ has only finitely many nilpotent orbits. Moreover, for all $x \in \mathfrak{h}$, we have* $\operatorname{Lie} C_H(x) = \mathfrak{c}_{\mathfrak{h}}(x)$.

The proof is essentially the same as that in 3.8, so we leave the details to the reader.

As in 3.9, case-by-case arguments for the simple groups (playing the role of H here) allow one to verify condition (*) whenever $\operatorname{char} K$ is good. In this situation, one concludes that *the Lie algebra has only finitely many nilpotent orbits.* Further case-by-case study (summarized in Carter [**29**, 5.11]) eventually extends this result to all characteristics. But there is no satisfactory conceptual argument comparable to the one given by Lusztig for the groups, which we discuss next.

3.11. Lusztig's Finiteness Theorem

Lusztig [**114**] proves in a uniform way:

THEOREM. *If G is simple and* $\operatorname{char} K = p > 0$, *then G has only finitely many unipotent classes.*

Coupled with either Richardson's Theorem 3.8 in good characteristic, or the classical work of Dynkin and Kostant in characteristic 0, this proves the finiteness theorem in full generality. Unfortunately, Lusztig's proof uses methods remote of $\hat{G}$ looks at the linear span of the restrictions to the unipotent subset in G^F of the irreducible characters involved. Using induction on the dimension of G, together with the Deligne–Lusztig technique of "induction" from a Levi subgroup to G^F, one argues that this is the space of all class functions on the unipotent set. Coupled with the previous bounds that depend only on G, this forces the finiteness of the number of unipotent classes in G.

Remark. As a byproduct of the reasoning just summarized, one sees that in a sense "most" centralizers of semisimple elements (in a group such as G^F where such centralizers are connected) [illegible] Levi factors of parabolic subgroups. This is [illegible]

CHAPTER 4

Regular Elements

As before, we denote by G a connected semisimple group of rank r; but everything we do can be generalized readily to the reductive case. An element $x \in G$ is called **regular** if $\dim C_G(x)$ is as small as possible, or equivalently (by 1.6 and 2.3), if $\dim C_G(x) = r$. We showed in 1.4 that the set G_{reg} of all regular elements in G is open (and therefore dense). When x is semisimple, we found a number of equivalent conditions for regularity in 2.3. We then showed in 2.5 that the regular semisimple elements already form a dense open subset of G. In this chapter we aim to study regular elements in more depth. This involves in particular showing that regular unipotent elements exist.

Almost everything in this chapter is drawn from the fundamental paper of Steinberg [**184**].

4.1. Unipotent Elements Lying in a Unique Borel Subgroup

We observed in 1.2 that the regular elements in $\mathrm{SL}(n, K)$ are the matrices whose characteristic and minimal polynomials coincide; in particular, the regular unipotent elements are all conjugate to the single Jordan block with 1's on and just above the diagonal. This suggests a reasonable guess about what such an element might look like in an arbitrary semisimple group G.

When we fix a Borel subgroup $B = TU$ and corresponding set $\{\alpha_1, \dots, \alpha_r\}$ of simple roots, we will denote by U_i the root group U_{α_i}. Given any total ordering of the positive roots, each $u \in U$ has a unique expression as a product of elements of the various U_α. Thanks to the commutation relations in U (0.5), the U_i-component of u is independent of the chosen ordering. We might guess that u will be regular just in case all its U_i-components are different from e. This eventually turns out to be true, but unfortunately the verification is not straightforward. Here are some preliminary observations.

PROPOSITION. *Let $u \in G$ be unipotent, and let n be the number (possibly infinite) of Borel subgroups containing u. The following conditions are equivalent:*

(i) $n = 1$.
(ii) $n < \infty$.

(iii) *If U is the unipotent radical of a Borel subgroup containing u, all U_i-components of u are different from e $(1 \leq i \leq r)$.*

Moreover, a regular unipotent element (if it exists!) must satisfy these conditions.

PROOF. Suppose (iii) holds, and write $B = TU$. Any other Borel subgroup B' containing u is conjugate to B. More precisely, using the Bruhat decomposition, we have $B' = vn_wBn_w^{-1}v^{-1}$ for some $w \in W$ and $v \in U \cap n_wU^-n_w^{-1}$ (see 0.7). Now B also contains $n_w^{-1}v^{-1}uvn_w$. Since $u^{-1}v^{-1}uv$ lies in the derived group of U, where all U_i-components are e, it follows that the U_i-component of $v^{-1}uv$ coincides with $u_i \neq e$. Conjugation by n_w^{-1} keeps this element in U, forcing $w^{-1}(\alpha_i) > 0$ for all simple roots α_i; this implies $w^{-1} = e = w$, so $n_w \in B$. Thus $B' = B$, and (i) holds.

On the other hand, if (iii) fails, then for some i the U_i-component of u is e. Thus u lies in the (infinitely many) distinct Borel subgroups $vn_iBn_i^{-1}v^{-1}$, where $n_i \in N$ represents the simple reflection $s_i = s_{\alpha_i}$ and $v \in U \cap n_iU^-n_i^{-1}$. So (ii) also fails. This implies the equivalence of (i), (ii), (iii).

Finally, suppose u is a regular unipotent element, but fails to satisfy (iii). Say the U_i-component $u_i = e$. This means that u lies in the unipotent radical $V = V_i$ of the minimal parabolic subgroup $P = P_i$ corresponding to the ith simple root. Now $\mathrm{Cl}_P(u) \subset V$. Since a Levi subgroup of P has semisimple rank 1, $\dim P - \dim V = r + 2$ (where $r = \operatorname{rank} G$). Thus

$$\dim C_P(u) = \dim P - \dim \mathrm{Cl}_P(u) \geq \dim P - \dim V > r,$$

contrary to the regularity of u in G. □

Remark. For later use, observe that this last step of the proof shows that $\dim C_G(u) \geq r + 2$ whenever u fails to satisfy conditions (i)–(iii). Replacing P by B in the calculation, we also deduce that $\dim C_B(u) \geq r + 1$.

The proposition tells us where to look for regular unipotent elements. There are plenty of elements satisfying (iii), but unfortunately it is very difficult to compute their centralizers directly and thus verify that they are actually regular. Even so, the equivalent conditions do provide some clues about the nature of the centralizer. If u lies in a unique Borel subgroup B, and $x \in C_G(u)$, then $u \in xBx^{-1}$. This forces $xBx^{-1} = B$, whence $x \in B$ (a Borel subgroup being equal to its normalizer by 0.3). Thus $C_G(u) = C_B(u)$. In turn, suppose $s \in B$ is a semisimple element which centralizes u, and let T be a maximal torus containing s, so $B = TU$. It follows from condition (iii) that s must satisfy $\alpha_i(s) = 1$ for all simple roots α_i. This forces $s \in Z(G)$, so finally $C_B(u) = C_U(u)Z(G)$.

4.2. Dimension of the Unipotent Variety

Denote by $\mathcal{U}$ the set of all unipotent elements of G. (We use a script letter to avoid conflict with the convention that U denotes a maximal unipotent subgroup

The strategy is to relate regular elements in G to regular unipotent elements in groups of smaller rank. If $x = su$ is the Jordan decomposition of an arbitrary element x, we know that $C_G(s)$ is reductive (2.2). Thus we can write $C_G(s)^\circ$ as the almost direct product $G'Z$, where the derived group G' is semisimple and Z is a central torus. From 1.12 we also know that u lies in $C_G(s)^\circ$ and therefore $u \in G'$. With $r = \operatorname{rank} G$, set $r' := \operatorname{rank} G'$. The key observation is as follows:

LEMMA. *Let $x = su \in G$, with $s = x_s$ and $u = x_u$. With notation as above, x is regular in G if and only if u is regular in G'. More precisely,* $\dim C_G(x) - r = \dim C_{G'}(u) - r'$.

PROOF. To compare dimensions, use the fact that $C_G(x)$ is the same as the centralizer of u in $C_G(s)$, which in turn is the almost direct product of Z and $C_{G'}(u)$. Thus $\dim C_G(x) = \dim C_{G'}(u) + \dim Z$. But $\dim Z = r - r'$, so $\dim C_G(x) - r = \dim C_{G'}(u) - r'$. □

Remark. For later use, observe that the proof of the lemma can be imitated to show that

$$\dim C_B(x) - r = \dim C_{B'}(u) - r',$$

if B' is a Borel subgroup of G' containing u and $B' \subset B$. Indeed, $C_B(x)$ is the same as the centralizer of u in $C_B(s)$; but $C_B(s) = B'Z$ because the left side is connected and solvable (see B10.6 or H19.4) and contains the Borel subgroup $B'Z$ of $C_G(s)$.

4.9. Borel Subgroups Containing a Regular Element

With Lemma 4.8 in mind, we want to compare the set $\mathcal{B}_x$ of Borel subgroups of G containing an arbitrary element $x = su$ with the set $\mathcal{B}'_u$ of Borel subgroups of G' containing u. As before, we write $C_G(s)^\circ = G'Z$, where G' is the derived group.

LEMMA. *Let $x = su$ be the Jordan decomposition of x in G.*

(i) *Each $B \in \mathcal{B}_x$ contains a unique element of $\mathcal{B}'_u$ (namely, $B \cap G'$).*

(ii) *Each element of $\mathcal{B}'_u$ is contained in at least one but at most finitely many elements of $\mathcal{B}_x$.*

PROOF. (i) Let $B \in \mathcal{B}_x$. Then s lies in some maximal torus $T \subset B$. In turn, B determines a set of positive roots in the set Φ of roots of G relative to T, and the unipotent radical U of B is the product of root subgroups U_α, $\alpha > 0$. From our study of centralizers of semisimple elements (2.2), we see that G' is generated by all root groups $U_\alpha \subset G$ for which $\alpha(s) = 1$. These α form the root system Φ' of G' relative to its maximal torus $T \cap G'$, and $B \cap G'$ is clearly the product of $T \cap G'$ together with all U_α $(\alpha > 0, \alpha \in \Phi')$. Thus $B \cap G'$ is a Borel subgroup of G'. Since $x \in B$, we have $u \in B \cap G'$ as required.

(ii) Let $B' \in \mathcal{B}'_u$. When will $B \in \mathcal{B}$ contain both x and B'? If this happens, $s \in B$ and in turn s lies in some maximal torus of B; this is also a maximal torus of $C_G(s)$ and therefore contains Z. So B must include the Borel subgroup $B'Z$ of $C_G(s)^\circ$. Conversely, if $B \in \mathcal{B}$ does include $B'Z$, then $x \in B$ follows because $u \in B'$ and the central element s of $C_G(s)^\circ$ lies in all its Borel subgroups (see H22.2, Corollary B).

Now $B'Z$ is a connected solvable group, so it lies in at least one Borel subgroup of G. On the other hand, $B'Z$ contains a maximal torus of $C_G(s)$ (which is also a maximal torus of G), and the number of $B \in \mathcal{B}$ containing this maximal torus is finite (equal to the order of the Weyl group W of G). So at most $|W|$ Borel subgroups contain $B'Z$. □

By combining the preceding two lemmas with Theorem 4.6, we reach the promised characterization of regular elements:

THEOREM. *An element of G is regular if and only if it lies in just a finite number of Borel subgroups. Moreover, if $x = su$ as above is regular and lies in n Borel subgroups of G, then*

(i) $n = |W|/|W'|$, *where W is the Weyl group of G and W' the Weyl group of G', the derived group of $C_G(s)^\circ$.*
(ii) $n = 1$ *if and only if u is regular in G and $s \in Z(G)$.*
(iii) $n = |W|$ *if and only if s is regular in G.*

PROOF. It remains to consider the number n when x is regular. In this case, u is regular in G' and therefore lies in a unique Borel subgroup of G'.

Fix a maximal torus T of $C := C_G(s)^\circ$ containing s. Then T lies in exactly $|W|$ Borel subgroups of G and $|W'|$ Borel subgroups of C. As in the above lemma, each of the former intersects C in one of the latter. By the conjugacy theorem for Borel subgroups of C, we see that $|W|/|W'|$ Borel subgroups of G containing T intersect C in a given Borel subgroup of C containing T. Precisely one of these is of the form $B'Z$, where B' is the Borel subgroup of G' containing u, so (i) follows.

Now $n = 1$ if and only if $W = W'$, or $G = G'$. This means $s \in Z(G)$, while u is regular in G. Finally, $n = |W|$ if and only if G' is trivial (forcing $u = e$), which means that s is a regular semisimple element of G. □

The theorem says that a regular element fixes only finitely many elements of the flag variety $\mathcal{B}$. But if x is irregular, the fixed point set $\mathcal{B}_x$ is much more difficult to describe as a variety. This becomes especially interesting in case x is unipotent: see Chapter 6 below.

4.10. Bijection Between Semisimple Classes and Regular Classes

The preceding analysis based on the Jordan decomposition makes it easy to derive the following basic relationship between semisimple and regular classes, which incorporates neatly the existence and conjugacy theorems for regular unipotent elements.

Having fixed the simple system Δ, we define for each distinct pair α and β of simple roots a nonnegative integer $c_{\alpha,\beta} := -\langle\beta,\alpha^\vee\rangle$. This can be 0,1,2, or 3. If $\operatorname{char} K = 2$ or 3, the definition must be modified to make the arguments below work in case roots α, β of two lengths occur and the ratio of the squared root lengths is $\operatorname{char} K$: in that case redefine $c_{\alpha,\beta} := 1$ if $-\langle\beta,\alpha^\vee\rangle = \operatorname{char} K$.

For distinct $\alpha, \beta \in \Delta$, we have to study closely the interaction of the minimal parabolic subgroups P_α, P_β and their unipotent radicals V_α, V_β. Note that $V_\alpha \cap V_\beta$ is normalized by B and is the product of root groups U_γ, where γ runs over all positive roots other than α, β. The lemma below involves, for a given $x \in V_\alpha$, choices of elements $y \in P_\alpha$ for which $y^{-1}xy \in V_\beta$. If y does this, then obviously so do all elements of the coset yB in P_α/B. Thus it makes sense to define

$$O_{\alpha,\beta} := \{x \in V_\alpha \mid \text{there are exactly } c_{\alpha,\beta} \text{ cosets } yB \in P_\alpha/B \text{ with } y^{-1}xy \in V_\beta\}.$$

LEMMA. *Let G be simple, with simple system Δ corresponding to a Borel subgroup B.*

(i) *Let $\alpha \neq \beta$ be simple roots belonging to adjacent vertices of the Dynkin diagram. For any $x \in V_\alpha$, there exists $y \in P_\alpha$ such that $y^{-1}xy \in V_\beta$.*

(ii) *Let $\alpha \neq \beta$ be simple roots. Then the set $O_{\alpha,\beta}$ defined above is a dense open subset of V_α.*

PROOF. All relevant calculations take place in $P_\alpha = L_\alpha V_\alpha$. By using convenient parametrizations $U_\gamma = \{x_\gamma(c)\}$ for the various root groups involved, one can write down explicit commutation formulas in B as well as formulas for the action of an element $n_\alpha \in P_\alpha$ representing the Weyl group reflection s_α. By starting with a Chevalley basis for a simple Lie algebra over $\mathbb{C}$, one insures that all relevant constants in such formulas are integers—though they may become 0 in fields of characteristic 2 or 3. (See Steinberg's lectures on Chevalley groups [**186**], as well as Carter [**26**, 5.2]. In H§33 some concrete formulas are written down in rank 2 cases).

The commutation in U can be quite complicated to work with, but fortunately a large part of U is irrelevant to the matter at hand. For brevity, denote the standard parabolic subgroup $P_{\{\alpha,\beta\}}$ by $P_{\alpha,\beta} = L_{\alpha,\beta}V_{\alpha,\beta}$. Here $V_{\alpha,\beta} \subset V_\alpha \cap V_\beta$ is the product of root groups U_γ for which γ is not a linear combination of α and β. Clearly $V_{\alpha,\beta}$ is normalized by both P_α and P_β. We only need to keep track of what is going on in the remaining factors $U_\gamma \subset L_{\alpha,\beta}$. The calculations below should be regarded as taking place in $V_\alpha/V_{\alpha,\beta}$, but to avoid putting bars over elements we abuse notation by just omitting the factors in $V_{\alpha,\beta}$. Thus the essential work can be done in the subsystem Ψ of Φ having basis $\{\alpha, \beta\}$.

From the Bruhat decomposition of P_α relative to B, we obtain coset representatives for P_α/B consisting of e together with all products $x_\alpha(c)n_\alpha$. Conjugating an element of $x \in V_\alpha$ by the inverse of one of these leads to explicit products of various $x_\gamma(a)$. In particular, n_α permutes the root groups just as s_α permutes the roots, with $x_\gamma(a)$ sent to $x_\delta(ma)$ for $\delta = s_\alpha\gamma$ and $m \in \mathbb{Z}$ depending on how n_α is chosen in N. For a single γ (but not all of them!) we can modify n_α in advance by a torus element to insure that $m = 1$. Consider the separate possibilities:

(1) Ψ is of type $A_1 \times A_1$. Here $c_{\alpha,\beta} = 0$, and only (ii) is at issue. Note that n_α and U_α centralize U_β. Thus for $y \in P_\alpha$, we get $y^{-1}xy \in V_\beta$ iff $x \in V_\beta$. This means $O_{\alpha,\beta} = V_\alpha - V_\beta$. This is a nonempty open subset consisting of elements with nonzero U_β-component.

(2) Ψ is of type A_2. Here $c_{\alpha,\beta} = 1$. Write a typical element of V_α in the form $x = x_\beta(a)x_{\alpha+\beta}(b)$. This lies in V_β just when $a = 0$. From the commutation relations we get $x_\alpha(c)^{-1}x_\beta(a)x_\alpha(c) = x_\beta(a)x_{\alpha+\beta}(ac)$, while $x_\alpha(c)$ commutes with $x_{\alpha+\beta}(b)$. On the other hand, the reflection s_α interchanges β and $\alpha+\beta$. As remarked above, the choice of n_α can be normalized so that $n_\alpha^{-1}x_{\alpha+\beta}(d)n_\alpha = x_\beta(d)$ for all $d \in K$. But $n_\alpha^{-1}x_\beta(a)n_\alpha = x_{\alpha+\beta}(ma)$ for some nonzero m depending on the choice of n_α. Combining these facts, we get

$$n_\alpha^{-1}x_\alpha(c)^{-1}x_\beta(a)x_{\alpha+\beta}(b)x_\alpha(c)n_\alpha = x_\beta(b+ac)x_{\alpha+\beta}(ma).$$

Now if $a = 0$, only the choice $y = e$ (or $y \in B$) gives $y^{-1}xy \in V_\beta$. If $a \neq 0$, the choice $y = x_\alpha(-b/a)n_\alpha$ is the unique one (up to right multiplication by B) for which $y^{-1}xy \in V_\beta$. Now (i) follows. For (ii), we claim that $O_{\alpha,\beta}$ is the nonempty open subset of V_α defined by the condition that either $a \neq 0$ or $b \neq 0$. In one direction, when $a \neq 0$, a unique coset yB (just described) yields $y^{-1}xy \in V_\beta$, while $a = 0$ and $b \neq 0$ affords the unique choice B. In the other direction, $a = b = 0$ means $x = e$, permitting every yB to conjugate x into V_β.

(3) Ψ is of type B_2, with α long and β short. Here $c_{\alpha,\beta} = 1$. We have $V_\alpha = U_\beta U_{\alpha+\beta} U_{\alpha+2\beta}$, and P_α normalizes $U_{\alpha+2\beta}$ (allowing us to ignore factors of this type). The computations modulo $U_{\alpha+2\beta}$ are essentially the same as those just done in type A_2.

(4) Ψ is of type B_2, with α short and β long. Here $c_{\alpha,\beta} = 2$ (unless $\operatorname{char} K = 2$, in which case the value is 1). We have $V_\alpha = U_\beta U_{\alpha+\beta} U_{2\alpha+\beta}$, with U_α centralizing $U_{2\alpha+\beta}$. Using appropriate parametrizations, we get:

$$x_\alpha(c)^{-1}x_\beta(a)x_\alpha(c) = x_\beta(a)x_{\alpha+\beta}(ac)x_{2\alpha+\beta}(ac^2),$$

$$x_\alpha(c)^{-1}x_{\alpha+\beta}(b)x_\alpha(c) = x_{\alpha+\beta}(b)x_{2\alpha+\beta}(2bc).$$

Combining these:

$$x_\alpha(c)^{-1}x_\beta(a)x_{\alpha+\beta}(b)x_{2\alpha+\beta}(d)x_\alpha(c) = x_\beta(a)x_{\alpha+\beta}(b+ac)x_{2\alpha+\beta}(ac^2+2bc+d).$$

Next, s_α fixes $\alpha+\beta$ and interchanges β and $2\alpha+\beta$. By choosing an appropriate representative n_α, we can arrange that $n_\alpha^{-1}x_{2\alpha+\beta}(f)n_\alpha = x_\beta(f)$. In this way, for $y := x_\alpha(c)n_\alpha$, the U_β-factor in $y^{-1}xy$ will be $x_\beta(ac^2+2bc+d)$.

Now part (i) of the lemma follows easily: If $a = 0$, choose $y = e$ (or any other $y \in B$). If $a \neq 0$, the quadratic polynomial $f(t) := at^2+2bt+d$ has at least one root $c \in K$, so we can choose $y := x_\alpha(c)n_\alpha$.

Part (ii) requires closer scrutiny of $f(t)$, whose discriminant is $\delta := b^2 - ad$. We claim that $x \in O_{\alpha,\beta}$ iff $\delta \neq 0$ (which clearly defines a nonempty open subset of V_α). Suppose $\operatorname{char} K \neq 2$. We have to look separately at the possibilities $a = 0$ and $a \neq 0$.

If $a = 0$, we are to show that $x \in O_{\alpha,\beta}$ iff $b \neq 0$. From the above calculations it is clear that precisely two cosets yB could produce $y^{-1}xy \in V_\beta$, namely the coset of e and the coset of $x_\alpha(-d/2b)n_\alpha$. The latter is defined iff $b \neq 0$. If $a \neq 0$, the quadratic polynomial $f(t)$ has two distinct roots (yielding two cosets yB) precisely when $\delta \neq 0$, proving our claim.

Finally, suppose $\operatorname{char} K = 2$. Here the only difference is that the quadratic polynomial in question has just one root, in agreement with the modified definition of $c_{\alpha,\beta}$.

(5) Ψ is of type G_2, with α long and β short. Here $c_{\alpha,\beta} = 1$. We have $V_\alpha = U_\beta U_{\alpha+\beta} U_{\alpha+2\beta} U_{\alpha+3\beta} U_{2\alpha+3\beta}$. But s_α fixes $\alpha + 2\beta$ and interchanges $\alpha + 3\beta$ and $2\alpha + 3\beta$, while conjugation by U_α stabilizes the product of these three root groups. Disregarding them, the required computations are just like those for type A_2 in case (2).

(6) Ψ is of type G_2, with α short and β long. Here $c_{\alpha,\beta} = 3$ (unless $\operatorname{char} K = 3$, in which case the value is 1). Now we have $V_\alpha = U_\beta U_{\alpha+\beta} U_{2\alpha+\beta} U_{3\alpha+\beta} U_{3\alpha+2\beta}$. Here the calculations are similar (but not identical) to those for type B_2 in case (4). They lead to a cubic polynomial, whose roots correlate with the $c_{\alpha,\beta}$ choices yB. As is traditional, we leave the details as a rainy-day exercise for the reader. □

Remark. Steinberg includes in his version of the above lemma a closer accounting of the possibilities for $x \in V_\alpha - O_{\alpha,\beta}$: either every yB can be used to conjugate x into V_β, or else there exists yB so that $z^{-1}y^{-1}xyz \in V_\alpha \cap V_\beta$ for all $zB \in P_\beta/B$.

5.9. Centralizer of a Subregular Element

When G is simple, we now know that the Richardson elements of minimal parabolic subgroups all lie in a single class, the subregular class of G. Its codimension is $r + 2$. What can be said about the centralizer of a typical subregular u? If u lies in a Borel subgroup B, with simple system Δ, we already know that $C_G(u)^\circ \subset P_\alpha$ for each $\alpha \in \Delta$. When $G = \mathrm{SL}(2, K)$, the only subregular unipotent is e, with centralizer $G = P_\alpha$. Leaving this extreme case aside, we can easily prove, in the spirit of Proposition 1.6:

PROPOSITION. *Let G be simple but not of type A_1. If u is a subregular unipotent element of G, then $C_G(u)^\circ$ is solvable.*

PROOF. Fix a simple root α, and say $u \in V_\alpha$. Abbreviate $V := V_\alpha$. Now $\mathrm{Cl}_V(u)u^{-1} \subset (V, V)$ implies $\dim \mathrm{Cl}_V(u) \leq \dim(V, V)$. In turn,

$$r + 2 = \dim C_G(u) \geq \dim C_V(u) \geq \dim V - \dim(V, V).$$

Here V is the product of root groups U_β, where β runs over all positive roots other than α. Since $r > 1$ and G is simple, it is clear that at least r of these do not occur in (V, V), corresponding to the $r-1$ simple roots $\beta \neq \alpha$ along with a root $\alpha+\beta$ (where β is a simple root not orthogonal to α). Thus $\dim V - \dim(V,V) \geq r$. Putting things together, we have $r+2 = \dim C_G(u) \geq \dim C_V(u) \geq r$. Since 3 is the smallest dimension of a semisimple group, we conclude that $C_G(u)^\circ$ coincides with its solvable radical. □

Case-by-case study eventually shows that $C_G(u)^\circ$ is *unipotent*—unless G is of type A or B, in which case a one-dimensional torus also occurs: see Steinberg [**188**, p. 155]. Further case-by-case work is required to pin down the component group $C_G(u)/C_G(u)^\circ$ of a subregular element (see 7.17 below). This also turns out to be solvable; for example, it is isomorphic to the symmetric group S_3 when G is of type G_2 (7.18). Thus $C_G(u)$ is always solvable when u is subregular and $r > 1$. (But for one unipotent class in the group of type E_8, the component group is the nonsolvable group S_5.)

5.10. Induced Unipotent Classes

To round out this chapter, we introduce briefly a far-reaching generalization of Richardson's work due to Lusztig and Spaltenstein [**123**]. The idea is easy to sketch, but unfortunately the proof of the theorem stated below involves some subtle dimension computations involving fixed point sets on the flag variety. These computations will be discussed in the next chapter, in the context of a desingularization of the unipotent variety. (For the Lie algebra analogue, see Collingwood and McGovern [**31**, Chap. 7], and for refinements in the case of nonconnected reductive groups, see Spaltenstein [**160**, II.3, II.7].)

Start with an arbitrary parabolic subgroup $P = LV$ of G, where L is a Levi subgroup and V is the unipotent radical. Let C be any unipotent class in L. Since V lies in all maximal unipotent subgroups of P, it is clear that the set CV consists of unipotent elements. As the image of $C \times V$ under the product map, this set is also irreducible. Because G has only finitely many unipotent classes, it follows that there is a unique unipotent class $\widetilde{C}$ in G which intersects CV in a (relatively) open dense subset. We call $\widetilde{C}$ the class **induced** by C. When $C = \{e\}$, $\widetilde{C}$ is the Richardson class in P.

Denote by $\mathcal{B}^G$, $\mathcal{B}^P$, $\mathcal{B}^L$ the respective flag varieties (collections of Borel subgroups) of G, P, L. Thus $\mathcal{B}^G \supset \mathcal{B}^P$, while the Borel subgroups of P are obtained by multiplying V with Borel subgroups of L. Fix an element $u \in C$ for which there exists $v \in \widetilde{C} \cap uV$, and use a subscript u or v to denote the fixed points on a flag variety: the set of Borel subgroups containing the given unipotent element. Evidently $\mathcal{B}^P_v \subset \mathcal{B}^G_v$, while $\mathcal{B}^P_v$ is in natural bijection with $\mathcal{B}^L_u$.

THEOREM. *With the above notation, we have:*

(i) $\dim C_G(v) = \dim C_L(u)$.

(ii) $\dim \mathcal{B}^G_v = \dim \mathcal{B}^L_u$.

(iii) $\widetilde{C} \cap CV$ *is a single P-class.*
(iv) $C_G(v)^\circ \subset P$.

The dimension statements here are the crux of the matter. They depend on the dimension formula quoted in (3) of 5.1 above. When one studies fixed point sets on the flag variety in more detail, one also gets a refinement of part (ii) of the theorem: $\mathcal{B}_v^P$ is a union of some of the irreducible components of $\mathcal{B}_v^G$.

While the construction of induced unipotent classes seems to depend strongly on the original choice of P, it turns out to depend only on L: associated parabolic subgroups yield the same induced classes, generalizing the Johnston–Richardson result. This requires a rather subtle proof.

CHAPTER 6

The Unipotent Variety and the Flag Variety

The main theme of this chapter is the relationship between the unipotent variety $\mathcal{U}$, on the one hand, and the flag variety $\mathcal{B}$, on the other. The fixed point varieties $\mathcal{B}_u$ are of special interest, since they occur as fibres of a natural desingularization of $\mathcal{U}$. While most of the results are valid for arbitrary G and K, the isogeny type of G sometimes makes a difference, and there are anomalies when char K is bad.

We continue to assume that G is a connected semisimple group, of rank r. Everything we do adapts readily to the case of a connected reductive group. But for nonconnected groups (such as the full orthogonal group) there are extra technical difficulties, addressed by Spaltenstein [**160**].

6.1. The Unipotent Variety

First we summarize briefly what we already know about the unipotent variety $\mathcal{U}$. By an indirect argument in 4.2, we found that $\mathcal{U}$ is a closed irreducible subset of G having codimension r. Thanks to 3.9 and 3.11, $\mathcal{U}$ is a union of finitely many G-orbits, the only closed one being $\{e\}$. There is a unique open dense orbit, consisting of regular unipotent elements (4.6). This may also be characterized as the set of simple points of $\mathcal{U}$ (4.24): from the perspective of 4.14, $\mathcal{U}$ is a single fibre of the Steinberg map.

Our first priority in this chapter is to find a desingularization $X \to \mathcal{U}$. The construction used in 4.2 turns out to be just what we want—at least, when G is simply connected. The fibres of the map in question are the fixed point sets $\mathcal{B}_u$ of unipotent elements on the flag variety $\mathcal{B}$. Much of the chapter is devoted to describing these fibres in more detail: they are connected (6.5), with irreducible components all of the same dimension (6.6), and this dimension is given by a very simple formula (6.7), which we shall refer to as the **Dimension Formula**:

$$\dim C_G(u) = r + 2\dim \mathcal{B}_u.$$

An immediate consequence is that *all unipotent classes have even dimension.* Eventually this carries over to all classes and fixed point varieties. Unfortunately, the most accessible proof of the Dimension Formula depends on the Bala–Carter theory to be described in the next chapter. This theory breaks down for fields of small prime characteristic, even though the formula itself is now known to be true in general (Spaltenstein [**160**, Theorem II.10.15]).

In the special case of Richardson elements, a self-contained proof (6.10) can be given in the context of 6.8. This takes care of $\mathrm{SL}(n, K)$ in particular—but in this case one can study the fibres even more directly (6.15). Another case in which the fibres can be described in some detail is that of subregular unipotents (6.11).

To conclude the chapter, we compare the unipotent and nilpotent varieties when $\operatorname{char} K$ is good.

6.2. Isogenies and the Unipotent Variety

Before we can formulate the desingularization theorem, we must clarify the role played by the isogeny class of G (cf. 0.10). There are central isogenies $G_{\mathrm{sc}} \rightarrow G$ and $G \rightarrow G_{\mathrm{ad}}$, where G_{sc} is simply connected and G_{ad} is of adjoint type. Since the center of any semisimple (or reductive) group consists of semisimple elements, a central isogeny maps the unipotent variety of the first group bijectively onto the unipotent variety of the second (and clearly sends regular elements to regular elements). But this map is an isomorphism of varieties if and only if it is *separable*: see H4.6. While separability is automatic if $\operatorname{char} K = 0$, there are a few tricky situations in prime characteristic, as outlined in 0.13.

Our construction of a desingularization of $\mathcal{U}$ works for a simply connected group—or a group G for which $G_{\mathrm{sc}} \rightarrow G$ is separable (we say that G has a *separable universal covering*). Under this hypothesis it follows from Theorem 4.24 that the simple points of the variety $\mathcal{U}$ are precisely the regular elements. This may fail in other cases, as pointed out by Slodowy [**156**, 3.9]: When $\operatorname{char} K = 2$, the unipotent variety of $\mathrm{SL}(2, K)$ has singularities at irregular elements (but is reduced as a scheme). On the other hand, the unipotent variety of $\mathrm{PGL}(2, K)$ has a more complicated description as a zero set in affine space, and is nonreduced as a scheme; but its underlying reduced variety is nonsingular.

As Steinberg [**190**] points out, the arguments of Springer [**172**] are based on the assumption that G is of adjoint type, but must be modified in light of this possible anomaly involving simple points.

6.3. A Desingularization of the Unipotent Variety

Recall what is meant by a **desingularization** (or resolution of singularities) of an irreducible variety V. Denote by V_s the nonempty open subset of V consisting of simple points. Then a desingularization of V consists of a pair (X, π), where X is an irreducible smooth variety and $\pi : X \rightarrow V$ is an epimorphism, such that π maps $\pi^{-1}(V_s)$ isomorphically onto V_s. The definition does not imply any sort of uniqueness. But when V has singular points, comparison with the smooth variety X—if it exists—may provide some insight into the nature of these singularities.

In 4.2 we constructed a variety which will now be shown to be a desingularization of the unipotent variety $\mathcal{U}$, under the (mild) hypothesis that the isogeny from the universal covering group G_{sc} to G is *separable*.

The construction went as follows. Fix a Borel subgroup B, with unipotent radical U, so that G/B can be identified with the flag variety $\mathcal{B}$. Then define

$$X := \{(gB, x) \mid g^{-1}xg \in U\} \subset G/B \times G,$$

and let $\pi : X \to \mathcal{U}$ be the projection to the second factor. Since every unipotent element of G has a conjugate in U, π is surjective. We showed that X is a closed irreducible subset of the product, and that $\dim X = \dim \mathcal{U} = \dim G - r = 2 \dim U$. In the following section we shall prove:

THEOREM. *Assume G has a separable universal covering. Then the pair (X, π) defined above is a desingularization of the unipotent variety $\mathcal{U}$ of G.*

6.4. Proof of the Theorem

For the proof of Theorem 6.3, we follow Springer [**172**], as modified by Steinberg [**190**, §6] (see also [**188**, 3.9]). The steps are fairly straightforward when $\operatorname{char} K$ is good and G has no simple factor of type A.

First, we show that (without any special hypothesis on G) the irreducible variety X is *smooth*. Let U^- be the product of the root groups U_α with $\alpha < 0$, so U^-B is open in G: the big cell in the Bruhat decomposition (0.7). Its image in G/B is the open set U^-B/B, whose inverse image in X relative to the projection $X \to G/B$ is denoted O. Thus O is an open subset of X, consisting of all pairs (vB, vuv^{-1}) with $v \in U^-$ and $u \in U$. Observe that the map $U^- \times U \to O$ sending $(v, u) \mapsto (vB, vuv^{-1})$ has a well-defined inverse, since v is the sole element of U^- in vB. Thus O is isomorphic as a variety to $U^- \times U$, and is in particular smooth. Now G acts naturally on X, by $z \cdot (gB, x) \mapsto (zgB, zxz^{-1})$, and the G-translates of O cover X: if $(gB, x) \in X$, then $u := g^{-1}xg \in U$ by definition, so $(gB, x) = g \cdot (eB, u)$ with $(eB, u) \in O$. So all points of X are simple.

The assumption that G has a separable universal covering implies, by 6.2, that the set of simple points of $\mathcal{U}$ is precisely the set $\mathcal{U}_{\mathrm{reg}}$ of regular unipotent elements. By Theorem 4.6, these form a single class in G. When $u \in \mathcal{U}_{\mathrm{reg}}$, we know from Theorem 4.9 that $\mathcal{B}_u$ consists of just one element. So the restriction of π to the inverse image $X_s \subset X$ of $\mathcal{U}_{\mathrm{reg}}$ is bijective.

All that remains to show is that π maps X_s isomorphically onto $\mathcal{U}_{\mathrm{reg}}$. When $\operatorname{char} K = 0$, this follows because π is separable. When $\operatorname{char} K$ is a good prime and G has no simple factor of type A, the assertion is an easy consequence of Richardson's Theorem 3.8, whose corollary (together with 1.5 and 1.10) shows that the orbit $\mathcal{U}_{\mathrm{reg}}$ of a regular unipotent element u is isomorphic to the quotient variety $G/C_G(u)$. The universal mapping property of a quotient then provides the needed inverse map from $\mathcal{U}_{\mathrm{reg}}$ to X_s.

In the remaining cases, we have to look more closely at the adjoint action on the Lie algebra, following [**190**, 6.1]. We use in a crucial way the fact that the Lie algebra of a simply connected group is obtained by reducing modulo p a Chevalley $\mathbb{Z}$-form of the corresponding Lie algebra over $\mathbb{C}$ (0.12). Since $G_{\mathrm{sc}} \to G$ is separable, this Lie algebra is isomorphic to $\mathfrak{g}$. In particular, we can work with a standard basis in $\mathfrak{g} = \mathfrak{u}^- \oplus \mathfrak{t} \oplus \mathfrak{u}$, where $\mathfrak{u}^- = \operatorname{Lie} U^-$, $\mathfrak{t} = \operatorname{Lie} T$, $\mathfrak{u} = \operatorname{Lie} U$. Take root vectors $x_\alpha \in \mathfrak{u}$, $y_\alpha \in \mathfrak{u}^-$ (for $\alpha > 0$), together with r *independent* $h_i = [x_i, y_i] \in \mathfrak{t}$, where $x_i = x_{\alpha_i}$ and $y_i = y_{\alpha_i}$ ($\alpha_1, \dots, \alpha_r$ simple).

The problem is to show that $\pi : X_s \to \mathcal{U}_{\mathrm{reg}}$ is separable, for which it is enough to consider the restriction of π to the open set O. Identifying O as above with $U^- \times U$, the map becomes $\varphi : U^- \times U \to \mathcal{U}$, sending $(v, u) \mapsto vuv^{-1}$. Both sides have the same dimension: $\dim G - r$. To apply the differential criterion for separability, we have to show that $d\varphi$ is an isomorphism at some point. This will follow (for reasons of dimension) if we can show that $(d\varphi)_{(e,u)}$ is injective, where $(e, u) \in O$ with $u \in U$ regular. As in 3.8, we use standard formulas for differentials (see H10.1) to calculate:

$$(d\varphi)_{(e,u)}(y, x) = x + (\operatorname{Ad} u^{-1} - 1)(y),$$

for $y \in \mathfrak{u}^-$ and $x \in \mathfrak{u}$. The $\operatorname{Ad} u^{-1}$ is needed to translate the cited formula from the tangent space at u to the tangent space at e.

To verify injectivity, we may work with u rather than u^{-1} for convenience. So we assume $x + (\operatorname{Ad} u - 1)(y) = 0$ and try to deduce that $x = y = 0$. Since u is regular, $\operatorname{Ker}(\operatorname{Ad} u - 1) \subset \mathfrak{b}$ (thanks to Corollary 4.6), while $y \in \mathfrak{u}^-$; so it suffices to show that $x = 0 = (\operatorname{Ad} u - 1)(y)$. This will follow immediately from:

LEMMA. *Let $u \in U$ be regular, $y \in \mathfrak{u}^-$. Write $(\operatorname{Ad} u)(y) = y' + h' + x'$ in $\mathfrak{u}^- \oplus \mathfrak{t} \oplus \mathfrak{u}$. If $h' = 0$, then $x' = 0$.*

PROOF. The prototype for this occurs in $\mathrm{SL}(2, K)$. Working with the standard basis (h, x, y) of $\mathfrak{g}$, and taking

$$u := \begin{pmatrix} 1 & 1 \\ 0 & 1 \end{pmatrix},$$

we find $(\operatorname{Ad} u)(y) = y + h - x$.

In general, take u to be in standard form $u_r \cdots u_1$ as in 4.1, with $u_i \neq e$ in $U_i = U_{\alpha_i}$. It is clear that $(\operatorname{Ad} u_i)(y_\alpha) \in \mathfrak{u}^-$ if $\alpha < 0$ but $\alpha \neq -\alpha_i$, while $(\operatorname{Ad} u_i)(h) \in h + \mathfrak{u}$ for $h \in \mathfrak{t}$. Using our $\mathrm{SL}(2, K)$ calculation, $(\operatorname{Ad} u_i)(y_i) = y_i + h_i - x_i$. Now compute inductively the effect of $\operatorname{Ad} u$, starting with the given element $z^{(0)} := y \in \mathfrak{u}^-$. Write $z^{(i)} := (\operatorname{Ad} u_i)(z^{(i-1)}) = y^{(i)} + h^{(i)} + x^{(i)} \in \mathfrak{u}^- \oplus \mathfrak{t} \oplus \mathfrak{u}$. Thus $z^{(r)} = (\operatorname{Ad} u)(y)$, $h^{(r)} = h'$, $x^{(r)} = x'$. If c_i is the coefficient of y_i in $y^{(i-1)}$, we find:

$$h^{(i)} = h^{(i-1)} + c_i h_i,$$

$$x^{(i)} = (\operatorname{Ad} u_i)(x^{(i-1)}) + (\operatorname{Ad} u_i - 1)(h_i) - c_i x_i.$$

The first equation yields $h^{(r)} = \sum c_i h_i$, so the assumption $h' = h^{(r)} = 0$ forces all $c_i = 0$ (the h_i being independent). In turn, the first equation shows that all $h^{(i)} = 0$. Now the second equation forces (by induction on i) all $x^{(i)} = 0$. This yields $x' = x^{(r)} = 0$ as desired. $\square$

This completes the proof of Theorem 6.3. Note that the last part of the argument is valid for all K, but is needed only when $\operatorname{char} K$ is bad or G has a factor of type A. On the other hand, the independence of the h_i fails when $G_{\mathrm{sc}} \to G$ is inseparable.

6.5. Connectedness of the Fibres

We now examine more closely the fibres $\mathcal{B}_u$ occurring in the desingularization of the unipotent variety constructed in 6.3. It is natural to ask whether they are irreducible, and if not, what are the dimensions of their irreducible components?

As an easy first step, we show (following an argument of Tits given in Steinberg [**188**, 3.9]) that the fibres are always *connected.* More precisely, two Borel subgroups containing a unipotent element u can be connected in $\mathcal{B}_u$ by a sequence of intersecting projective lines constructed as follows: Fix a Borel subgroup B, and identify $\mathcal{B}$ with G/B. If $P \supset B$ is a minimal parabolic subgroup (having a single negative root $-\alpha$), then $P/B \cong \mathbb{P}^1$ is a projective line in G/B. More generally, we refer to any translate gP/B as a **line of type** α.

THEOREM. *If $u \in G$ is unipotent, then the variety $\mathcal{B}_u$ of Borel subgroups containing u is connected. In fact, any two points can be connected by a sequence of projective lines in $\mathcal{B}_u$ of various types α.*

PROOF. To prove that $\mathcal{B}_u$ is connected, we can fix a Borel subgroup $B = TU$ such that $u \in U$; then we must show how to connect B to a distinct Borel subgroup ${}^gB \in \mathcal{B}_u$ by a sequence of projective lines of various types in $\mathcal{B}_u$. Thanks to the Bruhat decomposition, $g \in BwB$ for a unique $w \neq e$ in W, and we may then assume that $g = bn$ with $b \in B$ and $n \in N(T)$ representing w. It is natural to proceed by induction on $\ell(w)$.

Suppose $w = s_\alpha$ for a simple root α (relative to B). Set $v := b^{-1}ub$, so $v \in U \cap {}^wU = U^{(w)}$, the unipotent radical of the minimal parabolic subgroup P_α with unique negative root $-\alpha$. In turn, u lies in the unipotent radical of P_α and hence in all Borel subgroups of P_α. These form a projective line in $\mathcal{B}_u$ of type α, containing B and ${}^gB$.

For the induction step, let $w = s_1 s_2 \cdots s_q$ be a reduced expression (with each s_i a simple reflection). Say s_q is the reflection with respect to α, so $w\alpha < 0$ and the one-dimensional unipotent subgroup U_α intersects $U^{(w)}$ trivially (see 0.7). As above, define $v := b^{-1}ub \in U^{(w)}$, which in turn lies in the unipotent radical of P_α, as does u. In particular, $u \in s_q B s_q$, and this Borel subgroup is connected to B by a line of type α in $\mathcal{B}_u$. The induction hypothesis then provides a sequence of lines from $s_q B s_q$ to ${}^gB$ as required. $\square$

Remark. This argument is concrete and elementary. A quicker, but more sophisticated, proof can be based on the Zariski connectedness theorem: see Springer [**172**, 1.5].

Once the connectedness theorem is proved, we can deduce from the following lemma the connectedness of fixed point sets on **generalized flag varieties**: homogeneous spaces G/P, where P is parabolic.

LEMMA. *Let $P \subset Q$ be parabolic subgroups containing a fixed Borel subgroup B. If $x \in G$ is arbitrary, the restriction of the natural epimorphism $G/P \to G/Q$ maps the space of fixed points $(G/P)_x$ onto the space of fixed points $(G/Q)_x$.*

PROOF. If x fixes $gQ \in G/Q$, then $g^{-1}xg \in Q$ and in turn $g^{-1}xg$ lies in some Borel subgroup of Q. All of these have the form zBz^{-1}, $z \in Q$. Thus $x \in gzB(gz)^{-1}$, so x fixes $gzB \in G/B$, a coset which maps onto $gzP \in G/P$ and then onto $gzQ = gQ \in G/Q$. □

6.6. Irreducible Components of the Fibres

We have just seen that, for any unipotent element $u \in G$, the subvariety $\mathcal{B}_u$ of $\mathcal{B}$ is connected. It need not be irreducible, however. Indeed, the study of the family of irreducible components of $\mathcal{B}_u$ turns out to be interesting as well as intricate. What we can show, following Spaltenstein [**158**], is that the components are of equal dimension. This is intimately related to the same property of the variety obtained by intersecting the class of u with the unipotent radical U of a fixed Borel subgroup B of G. Call the class C, and write $Z := C_G(u)$.

To relate the two varieties, we use the quotient maps $\pi_1 : G \to C$, defined by $\pi_1(g) = g^{-1}ug$, and $\pi_2 : G \to \mathcal{B}$, defined by $\pi_2(g) = {}^gB$. These have respective fibres Zg and gB. The key point is that the inverse image under π_1 of $C \cap U$ is the same as the inverse image under π_2 of $\mathcal{B}_u$; this follows at once from the definitions, as the reader should verify. Denote the common inverse image by Y. It is closed in G, because $\mathcal{B}_u$ is closed in $\mathcal{B}$. Since $\mathcal{B}_u$ is connected (by Theorem 6.5) and the fibres of π_2 are connected, it follows that Y is connected; in turn, we see that its image $C \cap U$ is connected.

One immediate byproduct of this setup is a comparison of dimensions, which will be useful later on. Computing $\dim Y$ in two ways, we get:

$$\dim(C \cap U) + \dim C_G(u) = \dim \mathcal{B}_u + \dim B. \tag{1}$$

Now we can state the equidimensionality theorem:

THEOREM. *Let C be the class of a unipotent element $u \in G$. Fix a Borel subgroup B of G, with unipotent radical U. Then $C \cap U$ is connected. Moreover:*

(i) *All irreducible components of $\mathcal{B}_u$ have the same dimension.*

(ii) *All irreducible components of $C \cap U$ have the same dimension.*

PROOF. We have already shown that $C \cap U$ is connected. Now the strategy is to deal with (i) and (ii) in tandem: they turn out to be equivalent. Unfortunately, a certain amount of notation is required to deal with all the components involved.

Denote by $\{C_i \mid i \in I\}$ the set of irreducible components of $C \cap U$, and by $\{X_j \mid j \in J\}$ the set of components of $\mathcal{B}_u$. With $Z := C_G(u)$ as above, set $A := Z/Z^\circ$. It is clear that the conjugation action of Z on $\mathcal{B}$ preserves $\mathcal{B}_u$ and permutes the X_j, while the connected group Z° stabilizes each component; thus A permutes the X_j.

Next consider the closed sets $Y_j := \pi_2^{-1}(X_j)$. Because X_j and the fibres ${}^gB$ of π_2 are irreducible, Y_j must be irreducible. Obviously no Y_j is contained in the union of the others, so $\{Y_j \mid j \in J\}$ must be precisely the irreducible components of Y. Now $ZY = Y$, so A also permutes the components Y_j of Y, and all Y_j in an orbit are sent by π_1 to the same image. Since π_1 is a quotient map, and the A-orbit of Y_j is a closed set as well as a union of fibres, its image $\pi_1(Y_j)$ is also closed (and irreducible), hence lies in some component C_i. But C_i is irreducible and is covered by the closed sets $\pi_1(Y_j)$, $j \in J$; so at least one Y_j must in fact map *onto* C_i.

If we define $D_i := \pi_1^{-1}(C_i)$, we see from this discussion that the irreducible components of the closed set D_i are just the sets Y_j lying in a single A-orbit and mapped by π_1 onto C_i. In particular, $\dim D_i = \dim Y_j$. Comparing the dimensions of fibres and images of π_1 and π_2, we conclude:

$$\dim C_i + \dim Z = \dim X_j + \dim B. \tag{2}$$

This shows that (i) and (ii) in the statement of the theorem are indeed equivalent.

To complete the proof, start with any component X_j of maximal dimension. From 6.5, we know that any point of $\mathcal{B}_u$ can be reached from X_j by a sequence of projective lines in $\mathcal{B}_u$ corresponding to flag varieties of minimal (semisimple rank 1) parabolic subgroups of G. Let L be such a line intersecting X_j. It clearly suffices to show that L lies in some component of $\mathcal{B}_u$ of maximal dimension.

Now we can transfer the problem to $C \cap U$. Starting with a Borel subgroup $B' \in L \cap X_j$, let $v = \pi_1(\pi_2^{-1}(B'))$. Thanks to equation (2), v lies in the component $C_i = \pi_1(Y_j)$ of maximal dimension, and we just have to see that $\pi_1(\pi_2^{-1}(L))$ lies in a component of $C \cap U$ of maximal dimension. Say L is the set of Borel subgroups of the minimal parabolic subgroup ${}^gP$, where $P \supset B$. Now P is the union of its Borel subgroups, and all of these are P-conjugate and contain v. So the set $\mathrm{Cl}_P(v)$ of all P-conjugates of v must lie in the unipotent radical U_P (which is the unipotent part of the intersection of Borel subgroups of P). Moreover, $\mathrm{Cl}_P(v) = \pi_1(\pi_2^{-1}(L))$.

Let V be an irreducible component of $C_i \cap U_P$ containing v. If $\mathrm{Cl}_P(V) = V$, then $\mathrm{Cl}_P(v) \subset C_i$ and we are done. Otherwise, note that $\dim U_P = \dim U - 1$ (by minimality of P), so $\dim V \geq \dim C_i - 1$, and the larger irreducible variety $\mathrm{Cl}_P(V)$ has at least the same dimension as C_i; so its closure in $C \cap U$ is the required component. □

6.7. The Dimension Formula

Examination of special cases led Springer and Steinberg to discover an elegant formula relating the dimension of $\mathcal{B}_u$ to the dimension of $C_G(u)$ for a unipotent element u (and eventually for an arbitrary element of G). We refer to this as the **Dimension Formula**:

$$\dim C_G(u) = r + 2\dim \mathcal{B}_u.$$

While this formula has by now been verified in all cases, using the work of a number of people on the parametrization of unipotent classes, the most elementary proof is rather roundabout and breaks down when char K is a bad prime.

In the following section we shall reduce the proof of the formula to a density condition which can be more readily verified in many cases. Here we point out some equivalent formulations, the last of which doesn't explicitly involve the flag variety (but is no easier to prove!). We are following Carter [**29**, 5.10.2] and Spaltenstein [**160**, II.2.6].

PROPOSITION. *Let u be a unipotent element, C its class, and U a maximal unipotent subgroup of G. The following dimension statements are all equivalent:*

(i) $\dim C_G(u) = r + 2\dim \mathcal{B}_u$;
(ii) $\frac{1}{2}\dim C = \dim U - \dim \mathcal{B}_u$;
(iii) $\dim(C \cap U) = \dim U - \dim \mathcal{B}_u$;
(iv) $\dim(C \cap U) = \frac{1}{2}\dim C$.

PROOF. The equivalence of (i) and (ii) follows from the fact that $\dim C_G(u) = \dim G - \dim C = r + 2\dim U - \dim C$. The equivalence of (ii), (iii), (iv) is an immediate consequence of equation (1) in 6.6 above. $\square$

Remark. Although we are assuming that G is semisimple, the Dimension Formula generalizes immediately to connected reductive groups: On the one hand, the flag variety of such a group can be canonically identified with the flag variety of the semisimple derived group. On the other hand, $\dim C_G(u) - \operatorname{rank} G$ is clearly the same whether G denotes a reductive group or its derived group. In 6.17 below we will show that the truth of the Dimension Formula for G and its connected reductive subgroups of maximal rank will imply the truth of the analogous formula for arbitrary elements of G.

6.8. The Density Condition

Following Steinberg [**190**] (see [**29**, 5.10]), we derive the formula for fibre dimensions from a density condition. (When G is reductive but not necessarily connected, the arguments have been extended by Spaltenstein [**160**, II.10.15].)

Fix a Borel subgroup B of G, with maximal torus T and unipotent radical U. For a fixed unipotent class C in G, consider the following **Density Condition**:

(1) $\qquad C \cap U \cap {}^wU$ is dense in $U \cap {}^wU$ for some $w \in W$.

Here ${}^wU$ is shorthand for $n_w U n_w^{-1}$, where n_w is any representative in $N(T)$ for $w \in W$. We also use the notation ${}^wB$. For convenience, write $U^{(w)} := U \cap {}^wU$ and $B^{(w)} := B \cap {}^wB$.

THEOREM. *Let $u \in G$ be unipotent, and let $\mathcal{B}_u$ be the variety of Borel subgroups containing u. Then $\dim C_G(u) \geq \operatorname{rank} G + 2 \dim \mathcal{B}_u$. Equality holds if the class C of u satisfies the Density Condition* (1). *As a consequence, all unipotent classes of G have even dimension.*

PROOF. We introduce an auxiliary variety

$$S := \{(B_1, B_2, v) \in \mathcal{B} \times \mathcal{B} \times C \mid v \in B_1 \cap B_2\}.$$

To see that S is a closed subset of the product, we use reasoning like that in 0.15. Look first at the morphism $G \times G \times C \to G \times G$ defined by $(x, y, u) \mapsto (x^{-1}ux, y^{-1}uy)$. The inverse image of $B \times B$ is closed, and maps onto S under the obvious morphism $G \times G \times C \to \mathcal{B} \times \mathcal{B} \times C$. Since the fibres involve full cosets xB, yB, we conclude that S is closed.

It is easy to compute the dimension of S. The projection $S \to C$ is surjective, and the fibre over $v \in C$ is clearly isomorphic to $\mathcal{B}_v \times \mathcal{B}_v$. But $\dim \mathcal{B}_v = \dim \mathcal{B}_u$, so

$$\dim S = \dim C + \dim(\mathcal{B}_u \times \mathcal{B}_u) = \dim C + 2 \dim \mathcal{B}_u. \tag{2}$$

Observe next that S can be partitioned in a natural way into finitely many locally closed subsets indexed by W: Given any triple $(B_1, B_2, v) \in S$, we know that $B_1 \cap B_2$ contains a maximal torus (0.8), which is conjugate (say by g) to T. Thus ${}^gB_1$ and ${}^gB_2$ are Borel subgroups containing T, hence can each be written as ${}^wB$ for a suitable element $w \in W$. After modifying g by an element of $N(T)$, we can arrange that $B_1 = {}^xB$ and $B_2 = {}^{xw}B$ for some $x \in G$ and $w \in W$. Thanks to the Bruhat decomposition, the choice of w here is clearly unique. So S is the disjoint union of the sets

$$S_w := \{(B_1, B_2, v) \in S \mid B_1 = {}^xB, B_2 = {}^{xw}B \text{ for some } x \in G\}.$$

It is of course possible that some of these sets are empty.

To see that S_w is locally closed, notice that G acts naturally on S (by conjugation). The preceding discussion makes it clear that the sets S_w are just the G-orbits. Now it follows that $\dim S \geq \dim S_w$ for all $w \in W$, with equality holding for some w. So we need to say something about $\dim S_w$.

Using the fact that $({}^xB, {}^{xw}B) = ({}^yB, {}^{yw}B)$ if and only if $xB^{(w)} = yB^{(w)}$, we can define a bijective morphism

$$\{(B_1, B_2) \in \mathcal{B} \times \mathcal{B} \mid B_1 = {}^xB,\ B_2 = {}^{xw}B \text{ for some } x \in G\}$$

by $(B_1, B_2) \mapsto xB^{(w)}$. This induces an injection $S_w \to G/B^{(w)} \times C$, sending $({}^xB, {}^{xw}B, v) \mapsto (xB^{(w)}, v)$.

Unless S_w is empty, it is clear that S_w projects *onto* the first factor $G/B^{(w)}$. A typical fibre over $xB^{(w)}$ is clearly isomorphic to $C \cap {}^xB^{(w)} = {}^x(C \cap B^{(w)}) = {}^x(C \cap U^{(w)})$. Therefore all fibres have dimension equal to $\dim(C \cap U^{(w)})$. So for nonempty S_w we get

$$\begin{aligned} \dim S_w &= \dim G/B^{(w)} + \dim(C \cap U^{(w)}) \\ &= \dim G - \dim B^{(w)} + \dim(C \cap U^{(w)}) \\ &= \dim G - \dim T - \dim U^{(w)} + \dim(C \cap U^{(w)}). \end{aligned}$$

Combining this with (2) and the fact that $\dim S \geq \dim S_w$, we get

$$\dim S = \dim C + 2 \dim \mathcal{B}_u \geq \dim G - \dim T - \dim U^{(w)} + \dim(C \cap U^{(w)})$$

whenever S_w is nonempty, with equality for some w. Thus:

$$(\dim G - \dim C_G(u)) + 2 \dim \mathcal{B}_u = \dim G - r - \dim U^{(w)} + \dim(C \cap U^{(w)}),$$

or

$$\dim C_G(u) + (\dim U^{(w)} - \dim(C \cap U^{(w)})) = r + \dim \mathcal{B}_u.$$

This implies the first statement of the theorem.

Assume finally that the Density Condition (1) is true, noting that it can be fulfilled only by an element $w \in W$ for which S_w is nonempty. This shows that $\dim C + 2 \dim \mathcal{B}_u = \dim G - \dim T$, or $\dim C_G(u) = r + 2 \dim \mathcal{B}_u$ as required. □

Remark. If $\mathcal{B}$ is identified with G/B for some fixed Borel subgroup B, and d_u is defined to be $(\dim C_G(u) - r)/2$, the inequality in the theorem may be restated as: $\dim(G/B)_u \leq d_u$. Thanks to Lemma 6.5, we also get $\dim(G/P)_u \leq d_u$ for any parabolic subgroup P containing B. Equality need not hold here (take $P = G$, for example). But when it does hold, Steinberg [**190**, §3] develops more refined information about the components of dimension d_u in $(G/P)_u$. This is done in a rather general framework, relative to pairs $P, Q \supset B$. The analogue of the variety S introduced above is the set of triples $(gP, hQ, v) \in G/P \times G/Q \times C$ for which $g^{-1}vg \in P$ and $h^{-1}vh \in Q$. The orbits of $C_G(u)$ on pairs of components of $(G/P)_u$ and $(G/Q)_u$ of dimension d_u are shown to be in bijection with certain double cosets in $W_P \backslash W / W_Q$ (where W_P and W_Q are respective Weyl groups of the Levi subgroups of P and Q). These double cosets correspond to elements $w \in W$ for which the class of u meets $P \cap {}^wQ$ in a dense subset. When $P = Q = B$, one recovers the above Density Condition.

6.9. Verifying the Density Condition

Given an arbitrary unipotent class C, how can one check the Density Condition (1) in 6.8? Apart from special cases like Richardson classes (to be discussed below), there is at present no easy argument available. The most elementary approach (which requires that char K be 0 or "sufficiently large") relies on the Bala–Carter Theorem (see [**6**] or Carter [**29**, 5.9–5.10]). This theorem will be formulated carefully in the next chapter (7.14), where we survey the methods used to parametrize unipotent classes. But here we only need to quote the salient part of the conclusion:

> *Assume that* $\operatorname{char} K = 0$ *or that* $\operatorname{char} K = p \geq 3(h-1)$, *where* h *is the Coxeter number. If* C *is a unipotent class in* G, *there exist a parabolic subgroup* $P = LV$ *of* G *and a parabolic subgroup* Q *of the Levi subgroup* L *for which* C *intersects the unipotent radical of* Q *in its Richardson class.*

Of course, the proof of the theorem involves specifying how to locate the appropriate pair (L, Q).

The restriction here on the characteristic of K results from the use of much of the classical Dynkin–Kostant method, which cannot be adapted to characteristic p unless p is large enough. But Pommerening [**134, 135**] was able to supplement the arguments of Bala–Carter and extend their result to all cases in which $\operatorname{char} K$ is *good*. In particular, it is sufficient to assume $p > 5$. To remove this last restriction requires more sophisticated techniques (Spaltenstein [**160**, II.10.15]).

At any rate, suppose we take the quoted statement for granted. Then we can argue easily that the Density Condition is satisfied. Start with a fixed Borel subgroup $B = TU$ and corresponding set of simple roots $\Delta = \{\alpha_1, \dots, \alpha_r\}$. Without loss of generality, we may assume that the Levi subgroup L and its parabolic subgroup Q given by Bala–Carter are standard relative to Δ. Label standard parabolics in G with letters $I, J, \dots$ indicating subsets of the index set $\{1, \dots, r\}$, so $P = P_I = L_I V_I$. In turn Q corresponds to some subset $J \subset I$: the simple roots of a Levi factor of Q are the $\pm\alpha_j$, $j \in J$. Thus the unipotent radical of Q is $L_I \cap V_J$, and C intersects it in a dense subset.

Our task is done if we can identify $L_I \cap V_J$ with $U \cap {}^wU$ for an appropriate $w \in W$. Each of the root systems Φ, Φ_I, Φ_J has a longest element relative to the simple system induced by Δ: call these (respectively) $w_\circ, w_I, w_J$. We claim that $w := w_J w_I w_\circ$ will do the trick.

The root groups U_α in $L_I \cap V_J$ belong to roots $\alpha > 0$ for which $\alpha \in \Phi_I - \Phi_J$. For such α, applying w_J must yield a root with an unchanged α_i-component for some $i \in I - J$, hence $w_J\alpha > 0$ and $w_I w_J \alpha < 0$. Going the other way, the condition $w_I w_J \alpha < 0$ for $\alpha \in \Phi^+$ forces $\alpha \in \Phi_I - \Phi_J$. Indeed, α must lie in Φ_I since otherwise its image under $w_I w_J$ would involve a simple root α_i, $i \notin I$ and would still be positive. In turn, $w_J \alpha$ must be positive in Φ_I in order to be sent by w_I to a negative root; so $\alpha \notin \Phi_J$.

The condition on $\alpha \in \Phi^+$ that $w_I w_J \alpha < 0$ translates next into the condition that $\alpha = w_J w_I w_\circ \beta = w\beta$ for some $\beta > 0$. This describes exactly those U_α in $U \cap {}^wU$, and we are done. □

6.10. Richardson Elements

Without invoking any of the Bala–Carter theory, it is easy to complete the calculation of $\dim \mathcal{B}_u$ when u is a Richardson element: an element of the unique dense P-orbit in the unipotent radical V of a parabolic subgroup P (5.3). Here we follow Steinberg [**188**, p. 139] and [**190**, 4.1–4.2].

From Theorem 6.8 we have the general inequality:

$$\dim C_G(u) \geq r + 2\dim \mathcal{B}_u,$$

with equality just when the Density Condition is satisfied. For a Richardson element this condition follows immediately from the observation that $V = U \cap {}^wU$ for an element $w \in W$ which lies in the Weyl group of P and whose inverse sends the positive roots to negative roots in a Levi factor of P while leaving all roots in V positive.

Alternatively, we can get the needed reverse inequality even more directly by finding a subset of $\mathcal{B}_u$ which has the correct dimension. All Borel subgroups of P contain V, and thus lie in $\mathcal{B}_u$. By identifying $\mathcal{B}$ with G/B for some fixed Borel subgroup $B \subset P$, we therefore identify P/B with a (closed irreducible) subset of $\mathcal{B}_u$ and obtain $\dim \mathcal{B}_u \geq \dim P/B$. Now a simple dimension comparison can be made:

$$r + 2\dim \mathcal{B}_u \geq r + 2\dim P/B = \dim P - \dim V = \dim C_P(u) = \dim C_G(u).$$

The last equality comes from (2) in 5.4. This argument shows moreover that the image of P/B in $\mathcal{B}_u$ is an irreducible component. (There may however be others.)

Using either of these arguments, we conclude:

THEOREM. *Whenever u is a Richardson element of some parabolic subgroup of G, the Dimension Formula is valid:* $\dim C_G(u) = r + 2\dim \mathcal{B}_u$. □

6.11. Subregular Elements and Dynkin Curves

As in 5.7, we assume here that G is a *simple* algebraic group. So far a few general facts have been established about the projective variety $\mathcal{B}_u$: it is connected, with irreducible components of equal dimension, and (with full verification yet to be completed) this dimension is correlated explicitly with $\dim C_G(u)$. To get a more complete picture, one would of course want to know what the components look like, how many there are, and how they intersect. When u is regular, $\mathcal{B}_u$ consists of a single point, so there is nothing more to be said. The next case of interest is that of *subregular* unipotents.

Recall from 5.7 that when G is simple there is a unique unipotent class of codimension $r + 2$ in G, containing the Richardson elements of the various minimal parabolic subgroups. Thanks to 6.10 above, $\dim \mathcal{B}_u = 1$ when u is subregular. Moreover, the proof of Theorem 6.5 shows that $\mathcal{B}_u$ is a finite union of projective lines of type α for various simple roots α relative to a fixed Borel subgroup B (if we identify $\mathcal{B}$ with G/B).

Using suggestions of Tits, Steinberg [**188**, 3.10] gives a very precise description of $\mathcal{B}_u$ in this case. (This plays a critical role in Slodowy's treatment of singularities [**156**]: see 6.14 below.)

We need to recall some notation introduced in 5.7 and then quote the key lemma proved in 5.8. Having fixed the simple system Δ, we define for each distinct pair α and β of simple roots a nonnegative integer $c_{\alpha,\beta} := -\langle\beta,\alpha^\vee\rangle$. This can be 0,1,2, or 3. If $\operatorname{char} K = 2$ or 3, this definition is modified in case roots α, β of two lengths occur and the ratio of the squared root lengths is $\operatorname{char} K$: in that case define $c_{\alpha,\beta} = 1$ if $-\langle\beta,\alpha^\vee\rangle = \operatorname{char} K$.

Now we define a **Dynkin curve** in G/B to be a nonempty closed subset which is a union of projective lines of various types $\alpha \in \Delta$, subject to the requirement that each line of type α meets precisely $c_{\alpha,\beta}$ lines of type β if $\alpha \neq \beta$. Note that a Dynkin curve—if it exists—must be connected, since we are assuming that the Dynkin diagram is connected. Moreover, an easy argument (using the connectedness of the Dynkin diagram and the absence of circuits) shows that the total number of lines of type α must equal the ratio (1,2, or 3) of the squared length of α to the squared length of a short simple root. So the Dynkin curve consists of r lines if there is only one root length, but will otherwise have multiple lines corresponding to the longer simple roots.

The reader should draw a few typical pictures. For example, a Dynkin curve for type A_r must be a union of r projective lines, one for each simple root. If these roots are numbered in the standard way, the line of type α_1 meets just the line of type α_2, which meets both the line of type α_1 and the line of type α_3, etc. As Steinberg points out, the pictures (apart from labeling by roots) will be identical for G_2 and D_4, for F_4 and E_6, and for similar pairings involving classical types. (This is explained geometrically by Slodowy [**156**, 6.2].)

The following section will be devoted to the proof of the basic theorem:

THEOREM. *Let G be simple, and fix a Borel subgroup B. If $u \in G$ is a subregular unipotent element, then $(G/B)_u$ is a Dynkin curve.*

6.12. Details of the Proof

In order to prove the theorem, we have to sort out the geometry of lines in G/B. What does it mean for a line of type $\alpha \in \Delta$ to lie in $(G/B)_u$? How do such lines intersect, and how many can there be?

LEMMA. *Let u be a subregular unipotent element, and let $P_\alpha = L_\alpha V_\alpha$ be a minimal parabolic subgroup for $\alpha \in \Delta$.*

(i) *A line gP_α/B of type α is fixed pointwise by u if and only if $u \in gV_\alpha g^{-1}$. In particular, P_α/B lies in $(G/B)_u$ if and only if $u \in V_\alpha$.*

(ii) *For $\alpha \neq \beta \in \Delta$, a line of type α and a line of type β in G/B can intersect in at most one point.*

(iii) *The lines in $(G/B)_u$ of type α are finite in number and permuted transitively by $C_G(u)$, acting by left translation on G/B. In particular, if P_α/B is the only line of type α fixed pointwise by $u \in V_\alpha$, then $C_G(u) \subset P_\alpha$.*

PROOF. (i) To say that u fixes gP/B pointwise is to say that for all $x \in P$, $ugxB = gxB$, or $g^{-1}ug \in xBx^{-1}$. But the intersection of P-conjugates of B is $Z(G)V_\alpha$ (apply H26.1 to P), so the condition translates into: $u \in gV_\alpha g^{-1}$.

(ii) Suppose the lines xP_α/B and yP_β/B have two points in common: $xp_iB = yq_iB$ $(i = 1,2)$, with $p_i \in P_\alpha$ and $q_i \in P_\beta$. Then a quick calculation shows that $p_2^{-1}p_1$ lies in P_β as well as P_α. Since $P_\alpha \cap P_\beta = B$, we see in turn that $xp_1B = xp_2B$ as required.

(iii) We know from Theorem 5.3 that u lies in only finitely many G-conjugates of V_α, and these are permuted transitively by the conjugation action of $C_G(u)$. It follows from (i) that the lines gP_α of type α in $(G/B)_u$ are in bijection with these conjugates of V_α, and are also permuted transitively by $C_G(u)$. If $u \in V_\alpha$ fixes pointwise no line of type α other than P_α/B, then $C_G(u) \subset N_G(V_\alpha) = P_\alpha$ (see 0.8). □

Using this lemma and Lemma 5.8, we can easily complete the proof of Theorem 6.11. We are given a Borel subgroup B and a subregular element $u \in G$, and want to show that $(G/B)_u$ is a Dynkin curve. Note that replacing u by a conjugate does not essentially change the problem: $(G/B)_{gug^{-1}} = g(G/B)_u$ is a Dynkin curve if and only if $(G/B)_u$ is.

Start with a fixed (but arbitrary) $\alpha \in \Delta$. If β is any other simple root, Lemma 5.8 produces a dense open subset $O_{\alpha,\beta}$ of V_α consisting of those elements x for which precisely $c_{\alpha,\beta}$ cosets $yB \in P_\alpha/B$ satisfy: $y^{-1}xy \in V_\beta$. The intersection of these sets for $\beta \neq \alpha$ is again open and dense in V_α. Since the subregular elements are also dense in V_α, we may assume that our element u lies in all $O_{\alpha,\beta}$.

Now we invoke part (i) of the above lemma repeatedly. In particular, since $u \in V_\alpha$, the line P_α/B of type α lies in $(G/B)_u$. We claim that the $c_{\alpha,\beta}$ cosets $yB \in P_\alpha/B$ for which $y^{-1}uy \in V_\beta$ are in natural bijection with the lines of type β in $(G/B)_u$ meeting P_α/B. Given yB, part (i) of the lemma (applied to β in place of α) says that $yP_\beta/B \subset (G/B)_u$, and certainly yB lies in both lines. Going the other way, suppose a line gP_β/B of type β lies in $(G/B)_u$: again by (i), this implies $g^{-1}ug \in V_\beta$. If the line meets P_α/B in a point $pB = gqB$, where $p \in P_\alpha, q \in Q_\beta$, then $y := gq \in P_\alpha$. Thus $y^{-1}uy = q^{-1}g^{-1}ugq \in V_\beta$ because $g^{-1}ug \in V_\beta$. We conclude that $gP_\beta/B = yP_\beta/B$ meets P_α/B in the point yB, while $yB \in P_\alpha/B$ satisfies the condition $y^{-1}uy \in V_\beta$.

It remains to "translate" this information from P_α/B to an arbitrary line gP_α/B of type α in $(G/B)_u$. Part (iii) of the above lemma allows us to assume that $g \in C_G(u)$. Now a line hP_β/B of type β intersects gP_α/B iff $g^{-1}hP_\beta/B$

intersects P_α/B. Moreover, since $gug^{-1} = u$, the criterion in (i) shows that $hP_\beta/B \subset (G/B)_u$ iff $g^{-1}hP_\beta/B \subset (G/B)_u$. It follows that gP_α/B meets exactly $c_{\alpha,\beta}$ lines of type β in $(G/B)_u$.

Since α was arbitrary, and fixed point varieties of subregular elements are all translates of one another, this argument completes the proof of the theorem. □

6.13. Standard Dynkin Curves

Having proved that $(G/B)_u$ is a Dynkin curve when u is subregular, we might ask whether G/B contains any Dynkin curves other than these. The answer is no:

THEOREM. *Let G be simple, with Borel subgroup B. Any two Dynkin curves in G/B are translates of each other.*

Steinberg [**188**, p. 149] gives a detailed outline of the proof, which we shall summarize briefly here. Initially it seems quite difficult to pin down Dynkin curves in G/B. The proof of Theorem 6.11 required a lot of details, including separate study of rank 2 cases in Lemma 5.8. We worked with well-chosen elements u, for which explicit but rather complicated descriptions could be given of the lines in $(G/B)_u$.

But now we can take advantage of the translation action by G on the collection of Dynkin curves. The idea is to specify (for each simple type) a **standard Dynkin curve**, and then to show in step-by-step fashion that an arbitrary Dynkin curve can be translated to this one. To see how this works, recall from the proof of Lemma 5.8 the easy case of type A_2, recast in the language of lines in 6.12. The chosen element u (there denoted x) lies in V_α, and thus the line P_α/B of type α lies in $(G/B)_u$. This line intersects a unique line yP_β/B of type β in $(G/B)_u$, where usually $y = x_\alpha(d)n_\alpha$ for $d \neq 0$. But once we have this pair of intersecting lines in G/B, we can translate both of them by $x_\alpha(-d)$ to get another Dynkin curve consisting of the lines P_α/B and $n_\alpha P_\beta/B$ (with point of intersection $n_\alpha B$). This is our standard Dynkin curve for A_2.

Starting with this standard curve D_0, how can we compare it with an arbitrary Dynkin curve D in G/B? The latter contains a unique line gP_α/B of type α, so it is harmless to translate by g^{-1} at the outset and assume that D shares the line P_α/B with D_0. Using the Bruhat decomposition as in the proof of Lemma 5.8, we argue that the curve of type β in D is of the form $x_\alpha(d)n_\alpha P_\beta/B$; so we can perform another translation of D to recover D_0.

The proof of the theorem is more complicated than this, of course: one has to start with an arbitrary simple G, specify a standard Dynkin curve, and then work from vertex to vertex along the Dynkin diagram (starting at one end) to translate a given curve into this one. We shall just indicate what the standard curves look like. Using the standard numbering of vertices corresponding to simple roots $\alpha_1, \dots, \alpha_r$, write P_i for the minimal parabolic subgroup corresponding to α_i,

n_i for a representative of the simple reflection s_i, and $x_i(c)$ for $x_{\alpha_i}(c)$. Then a standard Dynkin curve D_0 is constructed as follows.

(A_r, $r \geq 1$) Take P_1/B, $n_1 P_2/B$, $n_1 n_2 P_3 B, \ldots, n_1 \cdots n_{r-1} P_r/B$.

(B_r, $r \geq 2$) Take P_r/B, together with translates by n_r of the lines in the standard Dynkin curve of type A_{r-1} corresponding to $\{\alpha_1, \ldots, \alpha_{r-1}\}$ and translates of the same $r-1$ lines by $x_r(c)n_r$ (for any fixed nonzero c).

(C_r, $r \geq 3$) Start with lines in the standard Dynkin curve of type A_{r-1} corresponding to $\{\alpha_1, \ldots, \alpha_{r-1}\}$, then add $x_{r-1}(c)n_{r-1}P_r/B$ and $x_{r-1}(d)n_{r-1}P_r/B$ for nonzero $c \neq d$.

(D_r, $r \geq 4$) Take P_{r-2}/B, $x_{r-2}(c)n_{r-2}P_{r-1}/B$, and $x_{r-2}(d)n_{r-2}P_r/B$ for nonzero $c \neq d$, together with translates by n_{r-2} of the lines in the standard Dynkin curve of type A_{r-3} corresponding to $\{\alpha_1, \ldots, \alpha_{r-3}\}$.

(E_r, $r = 6, 7, 8$) Lines are chosen here essentially as in type D. Removing the branch point (numbered 4) leaves three subdiagrams of type A with vertex sets $\{1,3\}$, $\{2\}$, $\{5, \ldots, r\}$ and corresponding standard Dynkin curves D, D', D''. Now take the line P_4/B together with those in $n_4 D$, $x_4(c)n_4 D'$, $x_4(d)n_4 D''$, for nonzero $c \neq d$.

(F_4) Take $n_3 P_4/B$ and P_3/B, together with $x_3(c)n_3 P_2/B$, $x_3(d)n_3 P_2/B$, $x_3(c)n_3 n_2 P_1/B$, $x_3(d)n_3 n_2 P_1/B$, for nonzero $c \neq d$.

(G_2) Take P_1/B, $n_1 P_2/B$, $x_1(c)n_1 P_2/B$, $x_1(d)n_1 P_2/B$, for nonzero $c \neq d$.

6.14. Singularities

We have examined in some detail the subregular class and the associated fibres $\mathcal{B}_u$ in the desingularization $\pi : X \to \mathcal{U}$ of the unipotent variety. The geometry here is of special interest, because it affords a natural setting for the study of Kleinian singularities (or "rational double points") and their desingularizations. This was observed by Grothendieck and Brieskorn (see Steinberg [**188**, pp. 156–158] and [**193**]) and then worked out thoroughly by Slodowy [**155, 156**].

Kleinian singularities arise from the actions of finite subgroups of SU_2 on $\mathbb{C}^2$, or the closely related actions of finite subgroups of SO_3 on $\mathbb{R}^3$ coming from the action of SU_2 on the second symmetric power of $\mathbb{C}^2$ (of dimension 3). Besides cyclic and dihedral groups, three well-known symmetry groups occur here: tetrahedral, octahedral, and icosahedral. The quotient of $\mathbb{C}^2$ by the group action is a surface S with an isolated singularity at the origin. This is correlated with a Dynkin diagram of type A, D, or E (or the corresponding simple algebraic group). The cyclic group of order $r+1$ yields a singularity of type A_r and the dihedral group of order $2r$ leads to type D_{r+2}, while the other three groups correspond to E_6, E_7, E_8 respectively.

Briefly stated, a minimal desingularization of the surface S yields a fibre over the singular point consisting of curves whose intersection pattern is precisely that of the Dynkin curve with the same label. To realize this explicitly inside the simple algebraic group G of type A, D, or E, one constructs a transverse slice to the class of a subregular unipotent element u. The slice intersects $\mathcal{U}$ in a two-dimensional surface, with the Kleinian singularity occurring at u. At the same time one gets a desingularization by taking the inverse image of this surface under π. If G has two root lengths, the construction also yields a singularity of type A, D, or E in accordance with the discussion preceding Theorem 6.11 above. (See Slodowy [**156**, 6.4].)

Slodowy places all of this in the appropriate general setting of "deformations" of singularities relative to group actions. He works with a simple algebraic group G over K, where $\operatorname{char} K = 0$ or $p > 4h - 2$ (h being the Coxeter number of the Weyl group of G). The latter restriction is imposed in order to pass to the Lie algebra $\mathfrak{g}$, where a specific transverse slice to the subregular nilpotent orbit can be constructed. There is a natural analogue for $\mathfrak{g}$ of the adjoint quotient (or Steinberg map) studied in Chapter 3, mapping $\mathfrak{g}$ to the orbit space $\mathfrak{t}/W$. The restriction of this map to the transverse slice is then shown to be a "semiuniversal" $\mathbb{G}_m$-deformation of the singularity in question, where $\mathbb{G}_m$ is the multiplicative group ([**156**, 8.7]). This formulation captures in a precise way all of the features sketched above.

6.15. General and Special Linear Groups

In the case of a subregular element u, we have seen how to describe $\mathcal{B}_u$ in considerable detail. To do something comparable for other unipotent elements is a more challenging problem. In the classical groups, starting with $\mathrm{GL}(n, K)$ or $\mathrm{SL}(n, K)$, unipotent elements can be parametrized by partitions or pairs of partitions (see 7.11 below). This suggests a combinatorial approach.

Steinberg [**190**, §5] treats $\mathrm{SL}(n, K)$ directly, using some standard information about representations of the symmetric group S_n. Spaltenstein [**157**] and [**160**, II.5] gives a similar but more self-contained account, recovering in the process some of the S_n theory. He then goes on to treat other classical groups in a parallel way: see [**160**, II.6]. Here we sketch his methods in the case $G = \mathrm{GL}(n, K)$ (whose flag variety is identifiable with that of $\mathrm{SL}(n, K)$). Without presupposing the theorems proved earlier in this chapter and without any restriction on $\operatorname{char} K$, he is able to show that $\mathcal{B}_u$ is connected, with equidimensional irreducible components of dimension given by the Dimension Formula 6.7. In doing this he describes the components explicitly, thereby counting them.

Recall from 1.3 that unipotent classes in G are in natural bijection with partitions $\lambda = [\lambda_1, \lambda_2, \dots, \lambda_d]$ of n: a Jordan matrix in the class corresponding to λ has Jordan blocks of sizes $\lambda_1 \geq \lambda_2 \geq \cdots \geq \lambda_d > 0$, with 1's on the diagonal and the superdiagonal of each block. In that section we were already able to compute directly the dimension of the class and centralizer in terms of λ. But the correlation with fixed points in $\mathcal{B}$ takes more work.

As in 1.3, each partition λ determines a **Young diagram**: a left-aligned array of empty boxes, with λ_1 boxes in the first row, λ_2 in the second row, etc. By inserting the numbers $1, 2, \dots, n$ in the boxes we obtain a **tableau**, which is called **standard** just when the numbers increase across each row and down each column. For example, the partition $[1^n]$ belonging to the class of e yields a unique standard tableau, as does the conjugate partition $[n]$ belonging to the regular unipotent class. But the partition $[n-1, 1]$ belonging to the subregular class admits $n-1$ standard tableaux: each number between 2 and n may appear in the unique box in the second row. Denote by $\mathrm{St}(\lambda)$ the collection of all standard tableaux associated with the diagram of λ.

Now fix a unipotent element u, with associated partition λ. The idea is to define a surjective map from $\mathcal{B}_u$ to $\mathrm{St}(\lambda)$. The fibres of this map then partition $\mathcal{B}_u$, and their closures turn out to be precisely the irreducible components of $\mathcal{B}_u$. Note that this is consistent with the several examples just mentioned. To carry out this program, we revert to the original concrete notion of "flag": identify G with $\mathrm{GL}(V)$, and define a (full) **flag** in V to be an ascending chain of subspaces $0 = V_0 \subset V_1 \subset \cdots \subset V_n = V$ with $\dim V_i = i$. Write $F := (V_0, V_1, \dots, V_n)$. The obvious action of G on flags permutes them transitively, with Borel subgroups being the isotropy groups. Thus the collection $\mathcal{F}$ of all flags is in bijection with G/B for any fixed Borel subgroup B. Using this identification, $\mathcal{B}_u$ becomes the set $\mathcal{F}_u$ of flags fixed by u, where $u \cdot F = F$ means $u(V_i) = V_i$ for all i.

With this notation, we can define the desired map $\tau : \mathcal{F}_u \to \mathrm{St}(\lambda)$. If u fixes F, it induces a (unipotent) operator on V/F_{n-i} for $0 \le i \le n$, whose Young diagram is denoted $d_i(F)$. Here $d(F) := d_n(F)$ is the Young diagram of λ. It is easy to see that $d_{i-1}(F)$ is obtained from $d_i(F)$ simply by removing a box from the end of one row. At each step in this process, we place the integer i in the box to be removed, thus assigning to each box in $d(F)$ a distinct integer between 1 and n. This is the desired standard tableau $\tau(F)$.

Consider for example the subregular case. Here the partition $\lambda = [n-1, 1]$ corresponds to a u-stable decomposition $V = U^1 \oplus U^2$, with $\dim U^1 = n-1$ and $\dim U^2 = 1$. One type of flag F fixed by u can be constructed by taking $V_1 = U^2$ and filling in the other V_i by adding to U^2 the subspaces in any u-fixed flag of U^1. Now $V/V_1 \cong U^1$ as u-modules, so the associated partition is $[n-1]$: its Young diagram is obtained from that of λ by removing the single box in the second row, so that box receives the integer n. The resulting standard tableau $\tau(F)$ has the integers $1, \dots, n-1$ in the first row. (By taking V_1 instead to be the line in U^1 fixed by u, the reader should be able to construct other types of flags with a prescribed integer $2, \dots, n-1$ in the second row of $\tau(F)$.)

The inductive nature of the construction of τ suggests an inductive approach to the proof that the closures of its fibres are the irreducible components of $\mathcal{F}_u$. To carry this out, Spaltenstein [**160**, II.4] first explores (for any reductive group G) the fixed points of unipotents on G/P for parabolic subgroups P. This is too technical to summarize quickly, but the application to $\mathrm{GL}(n, K)$ is fairly straightforward since Levi factors of parabolics are products of general linear groups. To summarize:

THEOREM. *Let $G = \mathrm{GL}(n, K)$, with flag variety $\mathcal{F}$. Let u be a unipotent element of G with Jordan form determined by the partition λ of n. Then there is a surjective map τ from $\mathcal{F}_u$ onto the set $\mathrm{St}(\lambda)$ of standard tableaux of shape λ, such that the closures of the fibres of τ are distinct and equal to the irreducible components of $\mathcal{F}_u$.*

Over a field of characteristic 0, the irreducible representations of the symmetric group S_n are parametrized in a natural way by the partitions of n. Readers familiar with this theory will recall that the degree of the representation of S_n corresponding to λ is the number of standard tableaux of a given shape λ. That this number coincides with the number of irreducible components above is not accidental. Springer has shown in the case of $\mathrm{GL}(n, K)$ or $\mathrm{SL}(n, K)$ that the Weyl group S_n acts irreducibly on the top cohomology of $\mathcal{F}_u$, whose dimension is the number of components. In general, his construction of Weyl group representations leads to analogous but more complicated results: see Chapter 9 below.

6.16. Fibres of Dimension 2

We return to the general setting of semisimple groups. Apart from the regular and subregular cases, not much is known about the structure of $\mathcal{B}_u$ as a projective variety beyond the results on the dimension and the number of irreducible components. We have seen that $\mathcal{B}_u$ must be a union of projective lines, but this leaves open a lot of possibilities.

Building on the work of Spaltenstein and others, Lorist [**111**] provides more detailed information in case $\dim \mathcal{B}_u = 2$, when G is a simple algebraic group of type A, D, or E. His arguments involve $\mathfrak{g}$ rather than G, so they transfer automatically to G only when $\operatorname{char} K$ is good (but they work in all cases for $\mathfrak{g}$). Here $\mathcal{B}_x$ for a nilpotent element $x \in \mathfrak{g}$ is the variety of Borel subgroups whose Lie algebras contain x.

Identifying $\mathcal{B}$ with G/B for some fixed Borel subgroup B, the basic idea is to study the intersections of $(G/B)_x$ with Bruhat cells in G/B. This requires case-by-work with root systems and Weyl groups, and use of the technique of induction (5.10). But the end result is a fairly simple description (in the simply-laced cases): Each irreducible component of $(G/B)_x$ is the product of two projective lines or is a ruled surface over a projective line having invariant $e = 2$ (as defined in Hartshorne [**66**, Chap. V]). Moreover, Lorist is able to describe the geometric intersections of components.

6.17. The Dimension Formula in General

We have concentrated throughout this chapter on the behavior of unipotent elements. But it is natural to look more generally at the fixed point variety $\mathcal{B}_x$ for arbitrary $x \in G$. Here the Connectedness Theorem 6.5 fails right away—for example, in the case of a regular semisimple element s, where $\mathcal{B}_s$ has $|W|$ elements (Theorem 4.9). But the Dimension Formula 6.7 turns out to be true for all x. More precisely, once we know that the formula holds for G and its other connected reductive subgroups of maximal rank, we can deduce (following Steinberg [**190**, §6]) the analogous formula for all $x \in G$:

$$\dim C_G(x) = r + 2 \dim \mathcal{B}_x.$$

To this end, start with the Jordan decomposition $x = su$, and set $H := C_G(s)$. We may assume that s lies in a fixed maximal torus T of G, so $T \subset H$. We know from 1.12 and 2.2 that u lies in H° and that $H \subset H^\circ N$ (where as usual $N = N_G(T)$). Moreover, the Borel subgroups of H° are the intersections $H^\circ \cap B$, where B runs over the Borel subgroups of G which contain s. (This follows at once from Theorem 2.2, as in the proof of Lemma 4.9(i). It can also be deduced in a more general setting from B11.14 combined with B11.16.) Fix one such B, so that G/B is isomorphic to $\mathcal{B}$ and $H^\circ/(H^\circ \cap B)$ is isomorphic to the flag variety $\mathcal{B}'$ of the reductive group H°. We are assuming the truth of the Dimension Formula for H° (see the remark in 6.7):

$$\dim C_{H^\circ}(u) = r + 2 \dim \mathcal{B}'_u.$$

Now $C_H(u) = C_G(x)$, so the desired dimension formula for x boils down to the assertion that $\dim \mathcal{B}_x = \dim \mathcal{B}'_u$. To relate the two flag varieties, we need to look more closely at fixed points of s on $\mathcal{B}$:

LEMMA. *Let W' be the Weyl group of H° relative to T. Then:*

(i) $(G/B)_s = H^\circ W B/B$.

(ii) *The disjoint sets $C_w := H^\circ wB/B$, where w ranges over coset representatives of W/W', are the irreducible (and connected) components of $(G/B)_s$.*

(iii) $C_w \cong H^\circ/(H^\circ \cap {}^wB) \cong \mathcal{B}'$.

PROOF. In one direction, s fixes each coset hwB with $h \in H^\circ$, $w \in W$, since $shwB = hswB = hws'B = hwB$ (where s' is a conjugate of s in T). In the reverse direction, suppose s fixes gB. Then $g^{-1}sg \in B$ has a B-conjugate $x^{-1}g^{-1}sgx$ in T. But two conjugate elements of T must be W-conjugate, so $w^{-1}sw = x^{-1}g^{-1}sgx$, forcing $gxw^{-1} \in H$ and then $g \in HwB$. Since $H \subset H^\circ N$, we conclude $gB \in H^\circ WB/B$. So (i) follows.

The natural map from H° onto $C_w = H^\circ wB/B$ induces a bijection between $H^\circ/(H^\circ \cap {}^wB)$ and C_w. This is actually an isomorphism of varieties: the map is separable, because $\operatorname{Lie}(H^\circ \cap {}^wB) = \operatorname{Lie} H^\circ \cap \operatorname{Lie} {}^wB$ (resulting from the fact that both groups contain T). Moreover, it is an isomorphism of H°-homogeneous spaces. Since $H^\circ \cap {}^wB$ is a Borel subgroup of H°, the isomorphism with $\mathcal{B}'$ follows as well. This proves (iii).

In particular, we see that C_w is a complete variety, hence *closed* in G/B. The sets C_w (for w ranging over coset representatives of W/W') are clearly disjoint and irreducible. Since $(N \cap H^\circ)/T = W'$, these C_w also fill up $(G/B)_s$ thanks to (i). So (ii) follows. □

We remark that the lemma has a natural generalization in which B is replaced by an arbitrary parabolic subgroup containing it (see Steinberg [**190**, 6.2]).

With the lemma in hand, it is easy to complete the argument. The isomorphism in (iii) between C_w and $\mathcal{B}'$ is clearly H°-equivariant. Fixed points of x on G/B are the same thing as fixed points of u on $(G/B)_s$, and this fixed point variety is a finite (disjoint) union of copies of $\mathcal{B}'_u$. So the dimensions agree as required. Moreover, all irreducible components of $\mathcal{B}_x$ have the same dimension, since this is true for $\mathcal{B}'_u$ (6.6). To summarize:

THEOREM. (i) *For any $x \in G$, all irreducible components of $\mathcal{B}_x$ have the same dimension.*

(ii) *Assume the Dimension Formula is valid for G and its connected reductive subgroups of maximal rank. Then for all $x \in G$, $\dim C_G(x) = r + 2\dim \mathcal{B}_x$. As a consequence, all classes in G have even dimension.* □

Remark. In the same vein one can adapt the earlier construction 6.3 to obtain a desingularization of any Steinberg fibre F, assuming again that G has a separable universal covering. Let s be a fixed semisimple element lying in the unique closed class in F and define X to be the set of pairs $(gB, x) \in G/B \times F$ for which $g^{-1}xg \in sU$. Then the projection $X \to F$ gives a desingularization, with fibres of the type $(G/B)_x$. (See Steinberg [**190**, 6.4].)

6.18. Fixed Point Characterizations

Their study of fixed points on generalized flag varieties G/P led Herb and O'Brian [**67**], in response to conjectures of Lusztig, to extend Steinberg's characterization 4.9 of regular elements and to characterize semisimple and unipotent elements in a similar spirit. To get meaningful information from the action of x on G/P one obviously has to assume that P is a proper parabolic subgroup. More precisely, if G has a number of simple factors $G_1, \dots, G_n$ and a simple system $\Delta \subset \Phi$ is partitioned accordingly into $\Delta_1 \cup \dots \cup \Delta_n$, assume that P is a standard parabolic whose root system intersects each Δ_i in a proper subset: P is *proper in all factors of* G. It would of course be easier just to look at simple groups, but one has to deal also with centralizers of semisimple elements and their parabolic subgroups. So in fact it is most natural here to allow G to be any connected reductive group.

THEOREM. *Let G be a connected reductive group, with a parabolic subgroup P which is proper in all factors of G. Set $\mathcal{P} := G/P$. Let $x \in G$, with fixed point set $\mathcal{P}_x$ on $\mathcal{P}$.*

(i) *x is the product of a unipotent element and an element of $Z(G)$ if and only if $\mathcal{P}_x$ is connected.*

(ii) *x is regular if and only if $\mathcal{P}_x$ is finite.*

(iii) *x is semisimple if and only if $\mathcal{P}_x$ is smooth, with x acting transversally (as defined below).*

Note that it is essential to allow a factor from $Z(G)$ for the "if" part of (i), since elements of $Z(G)$ fix $\mathcal{P}$ pointwise. This need not be mentioned in (ii) or (iii), since the multiple of a regular or a semisimple element element by a central element is again regular or semisimple.

Without reproducing all the details of the proof, we can outline briefly the main ideas involved. When $P = B$ some of the key steps (occurring independently in [**190**]) have in fact been encountered already in Lemma 6.17, which generalizes naturally to the setting of G/P. Recall also Lemma 6.5, which says in particular that the natural map $(G/B)_x \to (G/P)_x$ is *surjective*. This takes care of the "only if" part in (i) and (ii): When x is the product of a unipotent element u and a central element, $(G/B)_x = (G/B)_u$ is connected by Theorem 6.5, forcing $(G/P)_x$ to be connected as well. When x is regular, Theorem 4.9 says that $(G/B)_x$ is finite, forcing $(G/P)_x$ to be finite.

For the "if" part of (i), write $x = su$ in $B = TU$, with $s \in T$, $u \in U$. Assuming that $\mathcal{P}_x$ is connected, it has to be shown that $s \in Z(G)$. Now $\mathcal{P}_x = \mathcal{P}_s \cap \mathcal{P}_u$. From the generalization of Lemma 6.17, we get a partition of $\mathcal{P}_s$ into its connected components, now indexed by double cosets $W'wW_\Gamma$ (if $\mathcal{P} = G/P$ with $P = P_\Gamma$ standard). Each of these components is itself isomorphic to a generalized flag variety for $C_G(s)^\circ$, and is stable under u since $u \in C_G(s)^\circ$; so the fixed point set of u is nonempty and connected, by the "only if" part of (i). Thus the number of connected components of $\mathcal{P}_x$ equals the number of components of $\mathcal{P}_s$. But $\mathcal{P}_x$ is connected. Close scrutiny of the double cosets in question then shows that s must lie in $Z(G)$.

Consider next the "if" part of (ii), in which $\mathcal{P}_x$ is assumed to be finite; we have to show that x is regular. Writing $x = su$ as before, we see from the proof of (i) that u fixes a single point in each connected component of $\mathcal{P}_s$—itself a generalized flag variety for $C_G(s)^\circ$. Now x is regular in G if and only if u is regular in the derived group of $C_G(s)^\circ$ (4.8). This allows us to reduce the problem to the case of a unipotent element u fixing a unique point of $\mathcal{P}$. To avoid technical details, say G is simple and $u \in P$ with P a proper standard parabolic subgroup. We have to show u is regular. Otherwise (by 4.12) u lies in the unipotent radical V_α of some minimal parabolic $P_\alpha \subset P$. Take a simple root β adjacent to α in the Dynkin diagram. By Lemma 5.8, there exists $g \in G$ such that $g^{-1}ug \in V_\beta$. Thus $u \in gV_\beta g^{-1} \subset gPg^{-1}$. Since u lies in a unique conjugate of P, $gPg^{-1} = P$. But

$N_G(P) = P$, so $g \in P$. Now the parabolic subgroup $Q := gP_\beta g^{-1}$ normalizes its unipotent radical $V := gV_\beta g^{-1}$. So if $q \in Q$, we get $u \in V = qVq^{-1} \subset qPq^{-1}$, forcing $qPq^{-1} = P$. This means $Q \subset N_G(P) = P$. Because $g \in P$, this in turn forces $P_\beta \subset P$. Proceeding along the Dynkin diagram in this way, we get all minimal parabolics included in P and arrive at the contradiction $P = G$.

Finally, consider (iii). This requires a definition. If X is a smooth variety (such as $\mathcal{P}$), and φ an endomorphism of X (such as $\operatorname{Ad} x$) with smooth fixed point variety X^φ, then φ induces an endomorphism φ_* of the normal bundle of X^φ in X. We say φ **acts transversally** if $1 - \varphi_*$ is an automorphism of the normal bundle.

Now begin with a semisimple element s. As before, $\mathcal{P}_s$ is a disjoint union of generalized flag varieties, and is therefore smooth. The transversality condition amounts to looking at the action of $\operatorname{Ad} s$ on a tangent space to $\mathcal{P}$, say at P. This is essentially the same as the action on the complement $\operatorname{Lie} V^-$ of $\operatorname{Lie} P$ in $\mathfrak{g}$, where V^- is the unipotent radical of the opposite parabolic. We may assume that s lies in a maximal torus T of P, making it easy to compute fixed points on $\operatorname{Lie} V^-$. The upshot is that the tangent space to $\mathcal{P}_s$ at P may be identified with the span in $\operatorname{Lie} V^-$ of those $\operatorname{Lie} U_{-\alpha}$ with $\alpha(s) = 1$. The operator $1 - (\operatorname{Ad} s)_*$ is identified with $1 - \operatorname{Ad} s$ acting on the span of the remaining $\operatorname{Lie} U_{-\alpha} \subset \operatorname{Lie} V^-$, and this is obviously an isomorphism.

In the reverse direction, given $x = su$ with $\mathcal{P}_x$ smooth, one argues that $u = e$ by observing that otherwise the nonzero nilpotent operator $1 - (\operatorname{Ad} u)_*$ would make $1 - (\operatorname{Ad} x)_*$ fail to be an isomorphism on a suitable sub-bundle of the normal bundle of $\mathcal{P}_x$.

6.19. The Nilpotent Variety

Denote by $\mathcal{N}$ the set of nilpotent elements in the Lie algebra $\mathfrak{g}$ of G. Most of the results proved earlier about $\mathcal{U}$ have natural parallels for $\mathcal{N}$. Indeed, the study of $\mathcal{N}$ came earlier in the case $K = \mathbb{C}$ (see Kostant [**97, 98**]) and served as a model for later work on the unipotent variety. Whenever $\operatorname{char} K$ is good for G, it turns out that there is a G-equivariant isomorphism from $\mathcal{N}$ onto $\mathcal{U}$. This is due to Springer [**172**], with refinements by Bardsley and Richardson [**7**]. In particular, nilpotent orbits are in bijection with unipotent classes, permitting one to work out a parametrization of the latter by Lie algebra methods (see Chapter 7).

Before we explain Springer's theorem, we have to develop some information about $\mathcal{N}$. The following theorem is valid in all characteristics, but unfortunately some parts of the proof require case-by-case study when $\operatorname{char} K$ is bad. For these we omit the details, referring instead to the literature. Define an element $x \in \mathfrak{g}$ to be **regular** if $\dim C_G(x) = r$. (Then $\dim \mathfrak{c}_\mathfrak{g}(x) \geq r$; when equality holds, Springer [**170**] calls x **smoothly regular**.)

THEOREM. *Let G be any connected semisimple group.*

(i) *Viewing* $\mathfrak{g}$ *as an affine space,* $\mathcal{N}$ *is a closed irreducible subset of codimension* r.
(ii) *The number of* $\operatorname{Ad} G$*-orbits in* $\mathcal{N}$ *is finite.*
(iii) *There is a unique dense orbit in* $\mathcal{N}$*, consisting of the regular nilpotent elements.*

PROOF. The proof of (i) is very similar to the proof of Theorem 4.2. After fixing a Borel subgroup B with unipotent radical U, we define an auxiliary variety $Y \subset G/B \times \mathfrak{g}$ by

$$Y := \{(gB, x) \mid (\operatorname{Ad} g)^{-1}(x) \in \mathfrak{u}\}.$$

Since each nilpotent orbit in $\mathfrak{g}$ intersects $\mathfrak{u}$, the projection σ to $\mathfrak{g}$ maps Y onto $\mathcal{N}$. This implies that $\mathcal{N}$ is closed (which is clear anyway) and irreducible. The dimension calculation requires inspection of the other projection $Y \to G/B$. We need to know that some fibre here is finite, i.e., that some nilpotent element x lies in only finitely many Borel subalgebras. This is weaker than the assertion that regular nilpotent elements exist, and follows from a modified version of the proof of Proposition 4.1. Here one takes $x \in \mathfrak{u}$ to be a sum of root vectors corresponding to the simple roots. Assume that x also lies in some $(\operatorname{Ad} g)(\mathfrak{b})$. Writing $g = vn_w$ as in 4.1, we have $(\operatorname{Ad} n_w^{-1} v^{-1})(x) \in \mathfrak{u}$. Now $(\operatorname{Ad} v^{-1})(x) \in \mathfrak{u}$ has the same components relative to simple roots as x. Since $\operatorname{Ad} n_w^{-1}$ keeps this element in $\mathfrak{u}$, we conclude that $w = e$ and $g \in B$ as desired.

(ii) When $\operatorname{char} K$ is good, it follows from Richardson's Theorem 3.10 that there are only finitely many nilpotent orbits in $\mathfrak{g}$. This is actually true for any K, but requires a lot of case-by-case study (summarized by Carter [**29**, 5.11]). Hesselink [**70**] treated the classical types in characteristic 2, while Stuhler [**194**] settled the case of type G_2. For other exceptional types, see Spaltenstein [**161, 163**] and Holt–Spaltenstein [**74**].

(iii) Just as in 4.3, the existence and conjugacy of regular nilpotent elements follow at once from the finiteness of the number of nilpotent orbits. Note that the arguments for G in 4.5 do not carry over to $\mathfrak{g}$, so one has to rely on the special case approach indicated in (ii) when $\operatorname{char} K$ is bad. □

Remark. Regular nilpotent elements and their centralizers are explored by Springer [**170**, §5], under the assumption that $\mathfrak{g}$ is obtainable by base change from a Chevalley $\mathbb{Z}$-form. This happens when G is simply connected, but may fail otherwise: see 0.13.

6.20. The Unipotent Variety and the Nilpotent Variety

In order to reduce problems about $\mathcal{U}$ to problems about $\mathcal{N}$, we want to find a G-equivariant isomorphism between these varieties. When $G = \mathrm{GL}(n, K)$ or $\mathrm{SL}(n, K)$, the inverse maps $u \mapsto u - 1$ and $x \mapsto 1 + x$ obviously meet our requirement for any K. (But for other simple groups of type A_{n-1}, the relationship between $\mathcal{U}$ and $\mathcal{N}$ is not so transparent if $\operatorname{char} K$ divides n.)

For classical groups of types B, C, D (with $\operatorname{char} K \neq 2$), there is also an explicit map from $\mathcal{U}$ to $\mathcal{N}$. This comes from a partially defined map on the set $\mathrm{M}(n, K)$ of $n \times n$ matrices, as follows. Whenever x is a square matrix for which $1+x$ is nonsingular, it makes sense to map x to $\gamma(x) := (1-x)(1+x)^{-1}$. A routine calculation shows that $\gamma(x)+1$ is also nonsingular, and that $\gamma^2(x) = x$. Thus γ is a morphism on the open subset of $\mathrm{M}(n, K)$ defined by $\det(1+x) \neq 0$. Moreover, $\gamma(gxg^{-1}) = g\gamma(x)g^{-1}$ for $g \in \mathrm{GL}(n, K)$. This map originates in an 1846 paper of Cayley and is known as the **Cayley transform** (see Weyl [**200**, II, §10]). In particular, when x is unipotent and $\operatorname{char} K \neq 2$, $\gamma(x)$ is the product of a nilpotent matrix and a commuting nonsingular matrix, hence nilpotent. Similarly, when x is nilpotent, $\gamma(x)$ is unipotent.

Now specialize to the classical groups $\mathrm{SO}(2r+1, K)$, $\mathrm{Sp}(2r, K)$, $\mathrm{SO}(2r, K)$, of respective types B_r, C_r, D_r. Some further routine checking shows that the Cayley transform takes a unipotent element in each group to a nilpotent element of the corresponding Lie algebra (consisting of skew transformations relative to the symmetric or skew form involved). For example, if x is both unipotent and orthogonal, then $\gamma(x)$ is skew-symmetric, and similarly in the other direction. Therefore γ induces the desired G-equivariant isomorphism between $\mathcal{U}$ and $\mathcal{N}$ in these cases (with $\operatorname{char} K \neq 2$).

When $K = \mathbb{C}$, so that G is a Lie group, the logarithm map $G \to \mathfrak{g}$ and the exponential map $\mathfrak{g} \to G$ (given by the usual infinite series) can be used for all semisimple groups to relate $\mathcal{U}$ and $\mathcal{N}$. On these subsets the infinite series become polynomials, and thus make sense whenever $\operatorname{char} K = 0$. But the problem is more subtle for exceptional groups in prime characteristic. Usually there is no obvious way to construct an explicit isomorphism of varieties, so one has to be content with an existence proof. Moreover, bad prime characteristics lead to genuine exceptions: here the number of unipotent classes sometimes fails to equal the number of nilpotent orbits.

In formulating the main theorem, we may as well limit our attention to the simple case. Under the added assumption that $\operatorname{char} K$ is good for G, the isogeny type of G makes no real difference except in type A.

THEOREM. *Let G be simple, with $\operatorname{char} K$ good for G. In type* A, *further assume that the covering map $\mathrm{SL}(n, K) \to G$ is separable. Then there exists a G-equivariant isomorphism between the unipotent variety $\mathcal{U}$ of G and the nilpotent variety $\mathcal{N}$ of $\mathfrak{g}$.*

This is stated in a slightly weaker form by Springer [**172**, Theorem 3.1]: G is assumed to be simply connected, with $\operatorname{char} K$ good, and the G-equivariant map is only asserted to be a *homeomorphism*. This is of course somewhat unnatural in the context of algebraic geometry, though adequate for applications. The proof given by Springer (outlined also in Springer–Steinberg [**180**, III, §3]) would yield an isomorphism if $\mathcal{N}$ were known to be *normal* (like $\mathcal{U}$).

Bardsley and Richardson [**7**, §9] provide a more general setting for the theorem, obtaining it as a corollary to their version of Luna's "slice theorem" in prime characteristic. They also use a result of Demazure [**36**] to prove the normality of $\mathcal{N}$ in all cases. Their approach uses fairly sophisticated algebraic geometry in the setting of group actions. We shall instead discuss Springer's approach (which also invokes some serious geometric facts), because it builds on a number of our previous steps.

6.21. Springer's Method

The proof of Springer's original version of Theorem 6.1 may be divided into two parts, both of which are somewhat delicate and indirect. The essential difficulties arise in case $\operatorname{char} K = p > 0$.

(1) Fix a Borel subgroup B, with unipotent radical U. The first goal is to prove the existence of a B-equivariant isomorphism of varieties $\lambda : U \to \mathfrak{u}$ (a sort of "logarithm" map). This is done by working first in characteristic 0 and then specializing to K.

One would prefer to work over $\mathbb{Z}$, taking advantage of the Chevalley group scheme underlying G. But there are anomalies for bad prime characteristics (for example, the centralizer of a regular unipotent element in an adjoint group is no longer connected). Instead Springer works over the localization R of $\mathbb{Z}$ at the product of bad primes (if any) for G. In this setting he adapts his detailed calculations in [**170**, §4], involving the Chevalley commutation formulas in U. Starting with the analogues U_0 and B_0 of U and B over $\mathbb{Z}$, he uses base change to get R-schemes $U_1 = U_0 \times_{\mathbb{Z}} R$, etc. Now the goal is to find a B_1-equivariant isomorphism of R-schemes from U_1 onto $\mathfrak{u}_1$.

Under our hypotheses (in simple types other than A_r) we can shift to an adjoint group. Here the centralizer of a regular unipotent element is connected in G. Starting with such an element in U_1 (whose components relative to simple roots are nontrivial), Springer places this element in a 1-dimensional additive subgroup and then locates a regular nilpotent element of $\mathfrak{u}_1$ having precisely the same centralizer. This insures that the respective B_1-orbits are isomorphic. By comparing the logarithm map $U_1 \to \mathfrak{u}_1$, one gets an extension of this isomorphism between orbits to a B_1-equivariant isomorphism of R-schemes $U_1 \to \mathfrak{u}_1$. This in turn specializes to K.

(2) The B-equivariant map $\lambda : U \to \mathfrak{u}$ cannot readily be extended to a G-equivariant map $\mathcal{U} \to \mathcal{N}$. Instead, we compare the auxiliary varieties $X \subset G/B \times G$ (6.3) and $Y \subset G/B \times \mathfrak{g}$ (6.19). If $(gB, x) \in X$, then by definition $g^{-1}xg \in U$; so it makes sense to apply λ to obtain an element $y \in \mathfrak{u}$. Now $\operatorname{Ad} g^{-1}$ maps $(\operatorname{Ad} g)(y)$ into $\mathfrak{u}$. Because λ is B-equivariant, the resulting element of $\mathfrak{u}$ depends only on the coset gB. Thus the pair $(gB, (\operatorname{Ad} g)\lambda(g^{-1}xg))$ lies in Y. In this way we obtain a map $\theta : X \to Y$, which is clearly a morphism. Since λ is an isomorphism, θ is also an isomorphism. It is easily checked to be

G-equivariant relative to the natural action of G on X (6.4) and the analogous action on Y.

Now the projections $\pi : X \to \mathcal{U}$ and $\sigma : Y \to \mathcal{N}$ are well-behaved maps (proper and birational), and the variety $\mathcal{U}$ is normal (4.24). Even if $\mathcal{N}$ is not yet known to be normal, one can use the "Stein factorization" to write σ as a composite of maps $\sigma' : Y \to \mathcal{N}'$ and $\nu : \mathcal{N}' \to \mathcal{N}$. Here $\mathcal{N}'$ is normal and affine, while ν is bijective. Moreover, G acts naturally on $\mathcal{N}'$.

The idea now is to show that θ induces a G-equivariant isomorphism between $\mathcal{U}$ and $\mathcal{N}'$, which in turn composes with ν to give a homeomorphism between $\mathcal{U}$ and $\mathcal{N}$. For this one invokes Grothendieck's connectedness theorem [**64**, III, 4.3.1]. This insures that π_* maps the structure sheaf $\mathcal{O}_X$ onto that of $\mathcal{U}$, thereby inducing an isomorphism of global sections: $\Gamma(X, \mathcal{O}_X) \cong \Gamma(\mathcal{U}, \mathcal{O}_\mathcal{U})$. Similarly, $\Gamma(Y, \mathcal{O}_Y) \cong \Gamma(\mathcal{N}', \mathcal{O}_{\mathcal{N}'})$. But θ induces an isomorphism $\Gamma(X, \mathcal{O}_X) \cong \Gamma(Y, \mathcal{O}_Y)$. Because affine varieties are determined by the global sections of their structure sheaves, we get $\mathcal{U} \cong \mathcal{N}'$ as desired.

6.22. Summary

Let us take stock of what we know at this point about unipotent classes in G and nilpotent orbits in $\mathfrak{g}$.

- There are only finitely many unipotent classes in G, as a consequence of Richardson's theorem when char K is good (3.9) and Lusztig's theorem in arbitrary prime characteristic (3.11). Richardson's arguments apply equally well to nilpotent orbits when char K is good, but otherwise the finiteness theorem requires *ad hoc* arguments.
- The unipotent variety $\mathcal{U}$ is an irreducible closed subset of codimension r in G (4.2).
- Regular unipotent elements exist (4.5) and form a single conjugacy class in G (4.6). If u is regular and lies in a maximal unipotent subgroup U, then $C_G(u) = C_U(u)Z(G)$ is abelian of dimension r, with $C_G(u)^\circ = C_U(u)^\circ$. If char K is good, $C_U(u)$ is connected, but otherwise u fails to lie in $C_G(u)^\circ$ (4.7).
- Each parabolic subgroup has a unique dense orbit in its unipotent radical, yielding a Richardson class of easily computable dimension (5.3). Borel subgroups yield the regular class, while minimal parabolics all yield the subregular class (5.7). Except for type A (5.5), not all unipotent classes need be Richardson classes. There is a parallel development in $\mathfrak{g}$.
- The unipotent variety, whose simple points are the regular unipotents, has a desingularization (6.3) with fibres isomorphic to the fixed point varieties $\mathcal{B}_u$ on the flag variety (assuming that G has a separable universal covering). For any semisimple G, the sets $\mathcal{B}_u$ are connected (6.5), with equidimensional irreducible components (6.6). There is a simple

Dimension Formula (6.7): $\dim C_G(u) = r + 2\dim \mathcal{B}_u$. But except in the case of Richardson elements, the most elementary proof of this formula depends on the Bala–Carter Theorem to be discussed in the next chapter and places some mild restrictions on char K.

- When G is simply connected and char K is good, there is a G-equivariant isomorphism between the unipotent variety $\mathcal{U}$ in G and the nilpotent variety $\mathcal{N}$ in $\mathfrak{g}$ (6.20), inducing a bijection between unipotent classes and nilpotent orbits.

What information is missing? Except in the case of $\mathrm{SL}(n, K)$, we have not yet found a systematic way to enumerate the finitely many unipotent classes or nilpotent orbits (these problems being equivalent when char K is good). Such an enumeration should also reveal the dimension of each class, the structure of centralizers, and the partial ordering of classes induced by inclusions of classes in the closures of others. All of this will be surveyed in the next chapter, beginning with the classical Lie algebra approach of Dynkin and Kostant.

CHAPTER 7

Nilpotent Orbits and Unipotent Classes

Our aim here is to give a detailed survey of the ways in which unipotent classes in G and nilpotent orbits in $\mathfrak{g}$ can be enumerated. Most of this may be done under the assumption that G is simple. The results (but not always the methods) turn out to be reasonably uniform for each simple type A–G provided the characteristic of K is 0 or a good prime (3.9). Almost everything we say will require at least this restriction on K.

7.1. Some Questions

In previous chapters we have studied some natural questions about unipotent classes and the unipotent variety. But so far we have sidestepped the problem of making explicit lists of the finitely many classes for each simple type. There is of course a parallel problem for nilpotent orbits in $\mathfrak{g}$. In more detail, we might ask for the following information:

- For each simple group, provide an algorithm for listing and labelling the unipotent classes (or nilpotent orbits) and computing the total number of them. If possible, exhibit explicit class representatives, say of the type $\prod_{\alpha>0} x_\alpha(c_\alpha)$.
- Determine the dimension of each class (or orbit), i.e., the codimension of a corresponding centralizer.
- Identify the Richardson classes (or, more generally, the induced classes in the sense of 5.10).
- Describe in detail the structure of each centralizer (in G or $\mathfrak{g}$), including the structure of the component groups such as $C_G(u)/C_G(u)^\circ$. Exhibit a Levi decomposition (if it exists, which is not automatic in prime characteristic), and describe a Levi factor or at least its root system.
- The collection of classes (or orbits) may be partially ordered relative to inclusion of one in the closure of another. Exhibit this partial ordering in terms of the explicit parametrization.

Among the main sources for this chapter are: Dynkin [**49**], Kostant [**97, 98**], Bourbaki [**24**, VIII, §11], Springer–Steinberg [**180**, III, §4], Spaltenstein [**160**],

Carter [**29**], Collingwood–McGovern [**31**].

7.2. An Overview of the Classification

According to Theorem 6.20, when char K is good for G and G is simply connected (or has a separable universal covering), there is a G-equivariant isomorphism between the unipotent variety $\mathcal{U}$ in G and the nilpotent variety $\mathcal{N}$ in $\mathfrak{g}$. This permits us to concentrate primarily on the classification of nilpotent orbits in $\mathfrak{g}$. Recall (3.9) that the bad characteristics are the primes $p = 2$ for types B, C, D, E, F, G, p =3 for types E, F, G, and $p = 5$ for type E_8. There one is forced to rely on *ad hoc* methods and case-by-case analysis: see Carter [**29**, 5.11] for a detailed summary of results.

For classical groups, the natural matrix realizations can be used to good advantage (without much reliance on Lie theory), as seen in the work of Wall [**198**]). But the exceptional types such as E_8 call for a more unified approach. When char $K = 0$ the first comprehensive classification of nilpotent orbits was achieved (modulo a few errors) by Dynkin [**49**], with further refinements by Kostant [**97**] and others. Here each orbit is labelled by placing an appropriate weight 0, 1, or 2 at each vertex of the Dynkin diagram of $\mathfrak{g}$. Originally K was assumed to be $\mathbb{C}$, but suitable adaptations of the arguments allow K to be any algebraically closed field of characteristic 0. A thorough account from the Lie algebra viewpoint, incorporating the later methods of Bala–Carter, is given by Collingwood–McGovern [**31**].

In fact, with enough care one can make the proofs work in prime characteristic, provided p is sufficiently large: say $p > 3h - 3$, where h is the Coxeter number of the Weyl group. Moreover, the approach developed by Bala–Carter [**5, 6**] leads to a more efficient working out of the lists of weighted Dynkin diagrams. This is explained in detail by Carter [**29**, Chap. 5], together with the requisite part of the classical theory adapted to fields of large enough prime characteristic. Eventually some supplementary arguments due to Pommerening [**134, 135**] show that the main result of Bala–Carter is in fact valid for all good characteristics.

Since Collingwood–McGovern and Carter provide reasonably self-contained accounts of most of this work, we limit ourselves here to a careful exposition of the results, with some discussion of the methods of proof. After laying out the Dynkin–Kostant theory, we pause to look at some special cases including types A–D and G_2. Then we describe the Bala–Carter theory. The chapter concludes with a survey of several related issues which involve further subtleties: the component groups $C_G(u)/C_G(u)^\circ$, the closure ordering of unipotent classes, a related duality phenomenon, and the geometry of the closures of classes.

It should be remarked that a more direct group-theoretic approach to the determination of unipotent classes is possible. The Bruhat decomposition in G permits one (in principle) to compute explicitly the centralizer of a unipotent element, written in its standard form $\prod_{\alpha>0} x_\alpha(c_\alpha)$ in a maximal unipotent subgroup U. If one can somehow assemble a list of candidates for class representatives, it may be possible to show directly that they are nonconjugate (for example, because their centralizers are not isomorphic) and that every element of U is conjugate to one of them. This approach has in fact been used effectively in prime characteristic to investigate the exceptional groups and their finite subgroups in tandem (to be discussed in the next chapter). The Dynkin classification may be invoked here to suggest plausible class representatives—even for small characteristics where the theory breaks down.

7.3. Standard Triples and Representations

The main tool in the classical work on nilpotent orbits is the representation theory of $\mathfrak{sl}(2, K)$ (or "$\mathfrak{sl}_2$ theory", for short), applied to the adjoint action on $\mathfrak{g}$ of a subalgebra isomorphic to $\mathfrak{sl}(2, K)$ which contains a given nonzero nilpotent element. We start with a review of the basic facts needed.

Take as standard basis of $\mathfrak{sl}(2, K)$ the three elements:

$$h := \begin{pmatrix} 1 & 0 \\ 0 & -1 \end{pmatrix}, \; x := \begin{pmatrix} 0 & 1 \\ 0 & 0 \end{pmatrix}, \; y := \begin{pmatrix} 0 & 0 \\ 1 & 0 \end{pmatrix},$$

with Lie brackets

$$[hx] = 2x, \; [hy] = -2y, \; [xy] = h.$$

When the Lie algebra $\mathfrak{g}$ of a semisimple group G contains three nonzero elements (h, x, y) satisfying these same commutation relations (hence spanning a subalgebra isomorphic to $\mathfrak{sl}(2, K)$), we call the ordered triple (h, x, y) a **standard triple**. (The letters e and f are often used in place of our x and y, but we have already reserved e for the identity element of G.)

The finite-dimensional representations of $\mathfrak{sl}(2, K)$ are well-known and easily described in characteristic 0 (see [**82**, §7]). First of all, they are completely reducible. In turn, there is (up to isomorphism) precisely one irreducible representation $\rho_d : \mathfrak{sl}(2, K) \to \mathfrak{gl}(L_d)$ for each nonnegative integer d, with $\dim L_d = d+1$. The integer d is viewed as the *highest weight*: the eigenvalues of $\rho_d(h)$ are $d, d-2, \ldots, -(d-2), -d$, with corresponding eigenspaces (weight spaces) of dimension 1. Here $\rho_d(x)$ annihilates any vector of weight d.

When $\operatorname{char} K = p > 0$, the complete reducibility theorem fails badly; moreover, the only irreducible representations are of the types $\rho_0, \ldots, \rho_{p-1}$. But an $\mathfrak{sl}(2, K)$-module whose dimension is "small" relative to p will still be completely reducible, with irreducible summands of the type just listed. In particular, if $\mathfrak{sl}(2, K)$ is embedded in a larger algebra $\mathfrak{g}$ whose dimension is small relative to p, we can decompose the adjoint action just as in the classical case.

7.4. Jacobson–Morozov Theorem

The starting point for the study of nilpotent orbits in $\mathfrak{g}$ is the theorem of Jacobson–Morozov:

THEOREM. *Assume that* char $K = 0$ *or that* char $K = p > 3(h - 1)$, *where* h *is the Coxeter number. Then each nonzero nilpotent element* $x \in \mathfrak{g}$ *belongs to a standard triple* (h, x, y).

The theorem was originally proved in characteristic 0, but close scrutiny of the proof shows that everything can be done in prime characteristic provided $p > 3(h - 1)$. The point is that one needs to know that the Killing form of $\mathfrak{g}$ is nondegenerate. Because the determinant of the form may involve quite a few primes (see 3.9), the stated restriction is unavoidable. (In 10.2 below we sketch some related work on embedding unipotent elements into rank 1 subgroups of G.)

7.5. Conjugacy of Standard Triples

In general, the standard triple (h, x, y) into which a nonzero nilpotent element $x \in \mathfrak{g}$ can be embedded fails to be unique: for example, the adjoint action of $C_G(x)$ leaves x fixed but usually changes h and y. Of course, we are only concerned with the G-orbit of x, allowing even more flexibility in the choice of a standard triple. So it is essential to pin down the extent of the possible choices. Arguments of Kostant and Mal'cev lead to the following precise conjugacy statements:

THEOREM. *With the same restrictions as before on* char K, *standard triples* (h, x, y) *and* (h', x', y') *in* $\mathfrak{g}$ *are conjugate under* G *in these situations:*

(i) (h, x, y) *is conjugate to* (h', x, y') *under a unipotent subgroup of* $C_G(x)$ *whose Lie algebra is* $\mathfrak{c}_{\mathfrak{g}}(x) \cap [\mathfrak{g}, x]$;

(ii) (h, x, y) *is conjugate to* (h, x', y') *under* $C_G(h)^\circ$.

It follows easily that there are bijections among the following three collections:

- Nonzero nilpotent orbits in $\mathfrak{g}$
- Conjugacy classes of standard triples in $\mathfrak{g}$
- G-orbits of semisimple elements $h \in \mathfrak{g}$ belonging to standard triples

Indeed, the Jacobson–Morozov Theorem together with part (i) of the theorem yields a well-defined bijection between nonzero nilpotent orbits and classes of standard triples. Next, conjugate triples yield conjugate semisimple elements. In the reverse direction, suppose h and h' are conjugate and belong to standard triples (h, x, y), (h', x', y'). Then (h, x, y) is conjugate to a standard triple (h', x'', y''), which in turn is conjugate to (h', x', y') by part (ii) of the theorem.

7.6. Weighted Dynkin Diagrams

In the classical (characteristic 0) theory, it is not yet known at this stage that the number of nilpotent orbits in $\mathfrak{g}$ is finite. The transition to conjugacy classes of standard triples, or orbits of associated semisimple elements, is not immediately enlightening on this point. But implicit in the preceding development is a further important bijection between nilpotent orbits and **weighted Dynkin diagrams**, obtained from the Dynkin diagram of $\mathfrak{g}$ by attaching nonnegative integers to the vertices. These integers are the values $\alpha(h)$ for roots α in a simple system Δ chosen to be compatible with a standard triple (h, x, y). It turns out that only the integers 0, 1, 2 can actually occur, making the finiteness assertion obvious.

In more detail, the argument goes as follows. The element h of a standard triple lies in some Cartan subalgebra $\mathfrak{t}$ of $\mathfrak{g}$ (traditionally denoted $\mathfrak{h}$), the Lie algebra of a maximal torus T of G. If Δ is a basis for the root system Φ of G relative to T, we may identify Φ with the root system of $\mathfrak{g}$ relative to $\mathfrak{t}$ and use the same notation Δ for the simple roots (though strictly speaking the roots of $\mathfrak{g}$ relative to $\mathfrak{t}$ are differentials of the roots of G relative to T). In case $\alpha(h)$ is a nonnegative integer for all $\alpha \in \Delta$, we say that h is **Δ-dominant**.

Say Δ is fixed. When h is part of a standard triple, the eigenvalues of $\operatorname{ad} h$ are integers. Now the standard facts about fundamental domains for the Weyl group insure that the adjoint action of an element of $N_G(T)$ representing a suitable Weyl group element will take $h \in \mathfrak{t}$ to an element at which all simple roots take nonegative integral values. In other words, h is conjugate to a Δ-dominant element of $\mathfrak{t}$. (Equally well, we could keep h fixed and replace Δ by a W-conjugate simple system Δ' so that h is Δ'-dominant. But in working out examples below, we may find it more convenient to keep Δ fixed while considering a number of nonconjugate standard triples.)

Now the key fact is:

PROPOSITION. *If the element h of a standard triple (h, x, y) is Δ-dominant for the simple system Δ, then $\alpha(h) \in \{0, 1, 2\}$ for all $\alpha \in \Delta$.*

The proof, based on $\mathfrak{sl}_2$ theory, is easy. To Δ corresponds a Borel subgroup $B = TU$ and the opposite Borel subgroup $B^- = TU^-$, with respective Lie algebras $\mathfrak{b} = \mathfrak{t} \oplus \mathfrak{u}$ and $\mathfrak{b}^- = \mathfrak{t} \oplus \mathfrak{u}^-$. We have $x \in \mathfrak{g}_2 \subset \mathfrak{u}$ and $y \in \mathfrak{g}_{-2} \subset \mathfrak{u}^-$. If $x_\alpha \in \mathfrak{g}_\alpha$ for $\alpha \in \Delta$, then $[x_\alpha, y] \in \mathfrak{b}^-$ since y is a sum of negative root vectors. If $[x_\alpha, y] \neq 0$, it lies in the $\alpha(h) - 2 \leq 0$ eigenspace of $\operatorname{ad} h$ in $\mathfrak{b}^-$, forcing $0 \leq \alpha(h) \leq 2$ as desired. Otherwise x_α centralizes y. But by using $\mathfrak{sl}_2$ theory, we see that $\mathfrak{c}_{\mathfrak{g}}(y)$ is the sum of lowest weight vectors and hence lies in the sum of the subspaces $\mathfrak{g}_i$ for $i \leq 0$. This forces $\alpha(h) \leq 0$, so $\alpha(h) = 0$.

Combining this proposition with the results of 7.5, we conclude:

COROLLARY. *There is a bijection between the collection of nonzero nilpotent orbits in $\mathfrak{g}$ and a subset of the 3^r possible weighted Dynkin diagrams. In particular, the number of nilpotent orbits is finite.*

It is conventional to extend the bijection to all orbits by associating to the zero orbit the weighted diagram with all vertices labelled 0. Of course, the task of enumerating the orbits is far from complete at this stage. There is in fact no straightforward way to decide whether a given weighted Dynkin diagram arises from some nilpotent orbit. And a comparison for $\mathfrak{sl}(n, K)$ between 3^{n-1} and the number of partitions of n makes it clear that relatively few weighted diagrams may in fact arise this way. But before we grapple with this question, we should pause to consider what further information may be extracted from the adjoint action on $\mathfrak{g}$ of a given standard triple.

7.7. Parabolics

Fix a standard triple (h, x, y) and assume that a simple system Δ has been chosen so that h is Δ-dominant. Let $\mathfrak{b}$ be the Borel subalgebra of $\mathfrak{g}$ determined by Δ. The eigenvalues of $\operatorname{ad} h$ on $\mathfrak{g}$ are integers, so we obtain a natural $\mathbb{Z}$-grading of $\mathfrak{g}$ by defining

$$\mathfrak{g}_i := \{z \in \mathfrak{g} \mid [hz] = iz\}.$$

It is easy to verify that $[\mathfrak{g}_i, \mathfrak{g}_j] \subset \mathfrak{g}_{i+j}$. Of course, $x \in \mathfrak{g}_2$ and $y \in \mathfrak{g}_{-2}$. Note that $\mathfrak{g}_0$ is just the centralizer of h, and is therefore a reductive subalgebra.

It is clear that $\mathfrak{p} := \oplus_{i \geq 0} \mathfrak{g}_i$ is a subalgebra containing $\mathfrak{b}$ (thus is a parabolic subalgebra). Moreover, $\mathfrak{p}$ has a natural Levi decomposition compatible with the grading: $\mathfrak{p} = \mathfrak{l} \oplus \mathfrak{v}$, where $\mathfrak{l} := \mathfrak{g}_0$ and $\mathfrak{v} := \oplus_{i>0} \mathfrak{g}_i$. By looking closely at the conjugacy results in 7.5, one can check that $\mathfrak{p}$ is in fact determined uniquely by x, independent of the choice of standard triple in which x is embedded as well as the choice of Δ. We refer to $\mathfrak{p}$ as the **canonical parabolic** associated with x. When convenient, similar language may be applied to the corresponding parabolic subgroup P of G. (In [**31**] the term **Jacobson–Morozov parabolic** is used.)

In this way we have in a sense reversed the strategy of Richardson in 5.3 for associating to each parabolic subgroup of G (or parabolic subalgebra of $\mathfrak{g}$) a unipotent class (or nilpotent orbit). But except in very special situations, these two correlations between parabolics and nilpotents or unipotents are essentially unrelated. Indeed, not every parabolic will arise as a canonical parabolic, and we already saw that not every unipotent class is a Richardson class. Examples discussed below should reinforce these points.

7.8. Centralizers and Orbit Dimensions

The canonical parabolic provides a convenient setting for the study of the centralizer of the nonzero nilpotent element x. From $\mathfrak{sl}_2$ theory, we see immediately that $\mathfrak{c} := \mathfrak{c}_{\mathfrak{g}}(x)$ is spanned by highest weight vectors for the adjoint action of the subalgebra $\mathfrak{a}$ spanned by h, x, y. Since the weights involved are dominant relative to Δ, it follows that $\mathfrak{c} \subset \mathfrak{p}$. Moreover, the above Levi decomposition $\mathfrak{p} = \mathfrak{l} \oplus \mathfrak{v}$ induces a Levi decomposition of $\mathfrak{c}$. This is best seen by looking at the centralizer of $\mathfrak{a}$, which is just the sum of the trivial representations occurring (those of highest weight 0) and is therefore equal to $\mathfrak{c} \cap \mathfrak{l}$. An argument based on the nondegeneracy of the Killing form shows that this subalgebra is *reductive*, leaving $\mathfrak{c} \cap \mathfrak{v}$ as the nilradical of $\mathfrak{c}$.

The dimensions involved here can be read off easily from the grading of $\mathfrak{g}$. Define $d_i := \dim \mathfrak{g}_i$. The dimension of $\mathfrak{c}$ is just the total number of independent highest weight vectors occurring in the adjoint action of $\mathfrak{a}$. Since the highest weights are limited to 0, 1, 2 (all vectors of weights 1 or 2 being highest weight vectors), and since each occurrence of ρ_2 involves a vector of weight 0 which is not killed by $\operatorname{ad} x$, we conclude:

THEOREM. *Given a standard triple (h, x, y) in $\mathfrak{g}$, with associated grading, define $d_i := \dim \mathfrak{g}_i$. Then $\dim \mathfrak{c} = (d_0 - d_2) + (d_1 + d_2)$. In particular, the dimension of the orbit of x is computable from the weighted Dynkin diagram.*

Note that the dimension sum here mirrors the intersections of $\mathfrak{c}$ with $\mathfrak{l}$ and $\mathfrak{v}$, which have respective dimensions $d_0 - d_2$ and $d_1 + d_2$.

Examples below for G_2 will illustrate how effectively all of this may be computed in individual cases, since the grading of $\mathfrak{g}$ is easy to read off from the weighted Dynkin diagram and a knowledge of the root system.

Not only is $\mathfrak{c}$ contained in $\mathfrak{p}$, but at the group level we have $C_G(x) \subset P$, where P is the parabolic subgroup of G having Lie algebra $\mathfrak{p}$. This is a stronger statement than we encountered in the other direction, going from parabolic subgroups to Richardson elements u of G (Theorem 5.3): there we could only prove an inclusion of the type $C_G(u)^\circ \subset P$. (Indeed, examples in type G_2 below will show clearly the discrepancy between the two settings.)

7.9. Even Orbits

With this much machinery in place, we can make some precise observations about nilpotent orbits of special types (see Collingwood–McGovern [**31**, Chap. 4]).

In the Lie algebra setting, Kostant [**97**] initiated the study of the **principal** (i.e., regular) nilpotent orbit, consisting of all the nilpotent elements whose centralizers have dimension r. (We prefer to use the term "regular" here, following Steinberg's later group-theoretic development. But the term "principal" is more standard in the Lie algebra literature.) In analogy with the case of regular unipotent elements, a nilpotent element lying in a given Borel subalgebra is regular just when its components corresponding to simple roots are nonzero. Taking for x the sum of simple root vectors in $\mathfrak{g}$, one easily embeds x in a standard triple whose h element visibly satisfies: $\alpha(h) = 2$ for all simple α. Thus the weighted Dynkin diagram of the regular orbit has the label 2 at every vertex, the opposite extreme from the diagram representing the 0 orbit. (The regular orbit is the unique one of largest dimension in the nilpotent variety. It can be shown [**31**, 4.3] that there is also a unique nonzero orbit of smallest possible dimension: the **minimal orbit**.)

With the regular orbit and 0 orbit in mind, we define a nilpotent orbit to be **even** if all vertices of its weighted Dynkin diagram have labels 0 or 2. It is equivalent to require that $\dim \mathfrak{g}_1 = 0$, or that $\dim \mathfrak{g}_i = 0$ for all odd i. The relationship between evenness and other conditions on orbits is sometimes quite subtle. For example, once the full list of nilpotent orbits is in hand, inspection shows that the subregular orbit is always even—*except* in type A_r when $r > 3$ is even. But one interesting implication always holds:

PROPOSITION. *Every even orbit is a Richardson orbit: each element in the orbit is in fact a Richardson element of its canonical parabolic subalgebra.*

It is easy to sketch the proof. Let (h, x, y) be a standard triple, and let $\mathfrak{p} = \oplus_{i \geq 0} \mathfrak{g}_i$ be the canonical parabolic subalgebra associated to x. The representation theory of $\mathfrak{sl}(2, K)$ shows that the linear map $\operatorname{ad} x : \mathfrak{g}_0 \longrightarrow \mathfrak{g}_2$ is surjective. In turn, $\operatorname{ad} x$ maps $\mathfrak{p}$ onto the subspace $\mathfrak{m} := \oplus_{i \geq 2} \mathfrak{g}_i$. If P is the parabolic subgroup of G having Lie algebra $\mathfrak{p}$, it follows that the P-orbit of x in $\mathfrak{m}$ is dense. Assuming that the orbit of x is even, the space $\mathfrak{m}$ is the full nilradical of $\mathfrak{p}$, and thus x is a Richardson element of $\mathfrak{p}$.

The converse of the proposition fails, however. In type A_r, we know that every orbit is a Richardson orbit; but the discussion in the next section allows one to exhibit examples of odd orbits as soon as $r > 2$.

7.10. Example: Type A

It is instructive to compare the theory just outlined with the more direct parametrization of nilpotent orbits in $\mathfrak{sl}(n, K)$ given by partitions of n. As in 1.3 we denote a d-part partition by $[\lambda_1, \dots, \lambda_d]$, with $\lambda_1 \geq \lambda_2 \geq \dots \geq \lambda_d > 0$ and $\sum \lambda_i = n$. In a case like $[2, 1, 1, 1]$ we abbreviate by writing $[2, 1^3]$. The corresponding nilpotent orbit is represented by the Jordan matrix with blocks of sizes $\lambda_i \times \lambda_i$, each having entries 1 on the superdiagonal and 0 elsewhere. The partition $[1^n]$ corresponds to the zero matrix, which we leave aside in the following discussion. We follow Collingwood–McGovern [**31**, 3.6].

A $\lambda \times \lambda$ nilpotent Jordan block may be regarded as representing the nilpositive element of the standard basis of $\mathfrak{sl}(2, K)$ in the irreducible representation $\rho_{\lambda-1}$ of highest weight $\lambda - 1$. Thus we may associate to the nonzero nilpotent Jordan matrix parametrized by $[\lambda_1, \dots, \lambda_d]$ a standard triple (h, x, y) in $\mathfrak{sl}(n, K)$ representing the standard basis of $\mathfrak{sl}(2, k)$ in the representation

$$\bigoplus_{1 \leq i \leq d} \rho_{\lambda_i - 1}.$$

In particular, h has eigenvalues

$$\lambda_1, \lambda_1 - 2, \dots, -(\lambda_1 - 2), -\lambda_1, \dots\dots, \lambda_d, \lambda_d - 2, \dots, -(\lambda_d - 2), -\lambda_d.$$

The Weyl group here is the symmetric group S_n, acting by permutations of the diagonal entries. The upper triangular Borel subalgebra determines the positive roots, all of which take nonnegative values at a matrix $\operatorname{diag}(c_1, \dots, c_n)$ precisely when $c_1 \geq c_2 \geq \dots \geq c_n$. There the simple root α_i $(1 \leq i \leq n-1)$ takes the value $c_i - c_{i+1}$. To obtain the weighted Dynkin diagram we therefore have to reorder the entries of h in descending order and take successive differences. At one extreme, the partition $[n]$ belongs to the regular nilpotent orbit and leads to the diagram with all weights equal to 2. (It is an interesting exercise to show that the collection of weighted Dynkin diagrams belonging to nilpotent orbits is stable under the symmetry of the underlying Dynkin diagram.)

As a concrete example, take $\mathfrak{g} = \mathfrak{sl}(4, K)$, with Dynkin diagram

$$\circ - \circ - \circ$$

There are five nilpotent orbits, corresponding to the five partitions of 4.

Consider the nilpotent orbit in $\mathfrak{g}$ corresponding to the partition $[3, 1]$. The two blocks in the Jordan form involve respective sets of eigenvalues $\{2, 0, -2\}$ and $\{0\}$. Writing these in descending order as $2, 0, 0, -2$ and taking successive differences, we arrive at the weighted Dynkin diagram:

$$\overset{2}{\circ} - \overset{0}{\circ} - \overset{2}{\circ}$$

In particular, the orbit is even. The grading of $\mathfrak{g}$ is easy to work out: here $(d_0, d_1, d_2) = (5, 0, 4)$. It follows readily that the canonical parabolic has dimension 10, with 5-dimensional Levi subalgebra (having semisimple part of type A_1). Moreover, the centralizer of our nilpotent element has dimension 5 (forcing the orbit to have dimension 10), with 4-dimensional nilradical. This is in fact the subregular orbit, intersecting the nilradical of its canonical parabolic densely.

On the other hand, the partition $[2, 1^2]$ leads to an odd orbit (the only one occurring here), which has dimension 6 and is in fact the minimal nonzero orbit. Its weighted Dynkin diagram is

$$\overset{1}{\circ} - \overset{0}{\circ} - \overset{1}{\circ}$$

Here the canonical parabolic again has dimension 10 (it coincides with the parabolic above). Our nilpotent orbit fails to be the Richardson orbit for this parabolic; instead, it occurs as the Richardson orbit in a maximal parabolic.

The reader should verify the remaining parts of the table below.

partition	Dynkin weights	orbit dim	$\dim \mathfrak{l}$	$\dim \mathfrak{v}$	$\dim \mathfrak{c} \cap \mathfrak{l}$	$\dim \mathfrak{c} \cap \mathfrak{v}$
$[1^4]$	0 0 0	0	15	0	15	0
$[2, 1^2]$	1 0 1	6	5	5	4	5
$[2^2]$	0 2 0	8	7	4	4	3
$[3, 1]$	2 0 2	10	5	5	1	4
$[4]$	2 2 2	12	3	6	0	3

TABLE. Nilpotent orbits in $\mathfrak{sl}(4, K)$

7.11. Example: Other Classical Types

For groups (or Lie algebras) of types B, C, D, the unipotent classes (or nilpotent orbits) can also be conveniently parametrized by partitions. Semisimple Lie algebras in characteristic 0 are treated in detail by Collingwood–McGovern [**31**, Chap. 5], who work out explicit standard triples in the classical Lie algebras. Semisimple groups in arbitrary good characteristic (here meaning different from 2) are discussed in Springer–Steinberg [**180**, IV, §2]. (See also the summary in Carter [**29**, 13.1] and the more general treatment of classical groups in Wall [**198**].)

Each group in question can be realized naturally as a subgroup of a suitable $\mathrm{SL}(n, K)$: here n is respectively $2r+1$, $2r$, $2r$ for G of type B_r, C_r, D_r. Unipotent classes in G may be obtained by intersecting G with suitable unipotent classes in $\mathrm{SL}(n, K)$, and similarly for nilpotent orbits. Theorem 3.8 predicts in this situation that a class in $\mathrm{SL}(n, K)$ will break up into only finitely many classes in G. But in fact the situation is much more favorable: with a mild exception in type D_r (r even), each class in $\mathrm{SL}(n, K)$ yields at most one class in G. So partitions of n can again be used to parametrize unipotent classes or nilpotent orbits—here we emphasize the latter. An equivalent parametrization by *pairs of partitions* (λ, μ) also arises naturally in some of the literature. Write $|\lambda| := \sum \lambda_i$ for the sum of all parts in a partition.

(B_r) Nilpotent orbits in $\mathfrak{so}(2r+1, K)$ correspond to partitions of $2r+1$ in which all *even* parts occur with even multiplicity, or to pairs of partitions (λ, μ) satisfying $2|\lambda| + |\mu| = 2r+1$, where the parts of μ are required to be odd and distinct. For example, when $r = 8$, the partition $[4^2, 3, 2^2, 1^2]$ of 17 corresponds to the pair $\lambda = [4, 2, 1]$, $\mu = [3]$.

(C_r) Nilpotent orbits in $\mathfrak{sp}(2r, K)$ correspond to partitions of $2r$ in which all *odd* parts occur with even multiplicity, or to pairs (λ, μ) satisfying $|\lambda| + |\mu| = r$, where the parts of μ are required to be distinct. For example, when $r = 8$, the partition $[4, 3^2, 2^2, 1^2]$ of 16 corresponds to the pair $\lambda = [3, 2, 1]$, $\mu = [2]$ (to get the parts of μ, take half of each even part which occurs an odd number of times in the original partition).

(D_r) Nilpotent orbits in $\mathfrak{so}(2r, K)$ correspond to partitions of $2r$ in which all *even* parts occur with even multiplicity; but each "very even" partition (having only even parts, with even multiplicity) comes from two orbits—possible only when r is even. (This complicates notation, but is not difficult to explain: the two $\mathrm{SO}(2r, K)$-orbits in question are actually conjugate under the larger disconnected group $\mathrm{O}(2r, K)$. It is easy to visualize the corresponding weighted Dynkin diagrams: they are interchanged by the graph automorphism switching the two vertices numbered $r-1$ and r.) Alternatively, use a pair (λ, μ) satisfying $2|\lambda| + |\mu| = 2r$, where the parts of μ are odd and distinct. (There are two orbits when μ is empty and all parts of λ are even.)

From these parametrizations one can in turn derive the weighted Dynkin diagrams, in much the same spirit as for type A. For example, in type C_r, each part ν in the given partition of $2r$ is the dimension of a representation of $\mathfrak{sl}(2, K)$, whose weights are $\nu - 1, \nu - 3, \ldots, -(\nu - 1)$, and the total number of such weights is $2r$ (the sum of the dimensions). Combine all these weights in a list, arrange them in descending order as $\gamma_1 \geq \gamma_2 \geq \ldots \geq \gamma_{2r}$ (so $\gamma_r = 0$ or 1), and take successive differences $\gamma_i - \gamma_{i+1}$ to obtain the labels for vertices $i = 1, \ldots, r-1$ of the Dynkin diagram; the distinguished vertex r (belonging to the long simple root) gets the label $2\gamma_r$.

Remark. For classical groups, Spaltenstein [**160**, II.7] describes in detail the process of inducing unipotent classes (5.10). Eventually this leads to an explicit algorithm [**160**, III.10.7] for deciding which unipotent classes are Richardson classes. (For exceptional groups, see the tables due to Elašvili in [**160**, II, App.].)

7.12. Example: G_2

While the classical groups (or their Lie algebras) can be studied independently of the Dynkin–Kostant theory, the exceptional types are far more resistant to direct attack. This has provided strong motivation for the development of more unified methods. As a warm-up for the Bala–Carter Theorem, we take a brief look at type G_2, where in good characteristic there turn out to be five nilpotent orbits (or unipotent classes). Only three of these are Richardson orbits: the 0 orbit, the subregular orbit, and the regular orbit, of respective dimensions 0, 10, 12. How are the remaining orbits to be located?

First we might ask how much help we get from the Dynkin approach. It is instructive to look closely at a couple of obvious standard triples in the Lie algebra $\mathfrak{g}$ of type G_2, whose Dynkin diagram is

Denote by α the short simple root and by β the long simple root. We have $(\alpha, \beta^\vee) = -1$ and $(\beta, \alpha^\vee) = -3$.

One standard triple consists of $(h_\alpha, x_\alpha, y_\alpha)$, where x_α spans $\mathfrak{g}_\alpha$ and y_α spans $\mathfrak{g}_{-\alpha}$. Let $\mathfrak{a}$ be the corresponding subalgebra isomorphic to $\mathfrak{sl}(2, K)$. The adjoint action of $\mathfrak{a}$ on $\mathfrak{g}$ yields immediately a direct sum of three copies of $L(0)$, two copies of $L(3)$, and one copy of $L(2)$ (this being $\mathfrak{a}$ itself). In particular, the respective dimensions of the $\mathfrak{g}_i$ $(-3 \leq i \leq 3)$ are $2, 1, 2, 4, 2, 1, 2$. Thus the canonical parabolic $\mathfrak{p}$ has a Levi subalgebra $\mathfrak{l}$ of dimension 4 (with derived algebra of type A_1, corresponding to a long root), and a nilradical $\mathfrak{v}$ of dimension 5. Indeed, it is the Lie algebra of a minimal parabolic subgroup.

The centralizer $\mathfrak{c}$ of x_α is also easy to locate: it is spanned by highest weight vectors in the six irreducible modules listed above, and has a Levi factor of dimension 3 (with a basis consisting of $h_\alpha + 2h_\beta$, $x_{3\alpha+2\beta}$, $y_{3\alpha+2\beta}$) and a nilradical also of dimension 3 (with a basis consisting of $x_{3\alpha+\beta}$, x_α, $x_{-\beta}$). Thus the G-orbit of x_α has dimension 8; in particular, it is not one of the Richardson orbits.

Note that $\alpha^\vee$ is the long simple root in $\Phi^\vee$, and is conjugate under the Weyl group to the highest root (a dominant weight) $2\alpha^\vee + 3\beta^\vee = (2\alpha+\beta)^\vee$. So if we replace h_α by $h_{2\alpha+\beta}$, we can read off the weighted Dynkin diagram: α and β get respective weights $\alpha(h_{2\alpha+\beta}) = \langle \alpha, (2\alpha+\beta)^\vee \rangle = 1$ and $\beta(h_{2\alpha+\beta}) = \langle \beta, (2\alpha+\beta)^\vee \rangle = 0$. This corresponds to the class name $\widetilde{\mathrm{A}}_1$ in the table below.

Encouraged by this success, we next consider the standard triple $(h_\beta, x_\beta, y_\beta)$ corresponding to the long simple root β. As before, it is easy to compute the adjoint action on $\mathfrak{g}$ of the resulting copy of $\mathfrak{sl}(2,K)$. We find a copy of $L(2)$, four copies of $L(1)$, and three copies of $L(0)$, leading to subspaces $\mathfrak{g}_i$ $(-2 \le i \le 2)$ of respective dimensions $1,4,4,4,1$. As a result, the canonical parabolic has a Levi subalgebra of dimension 4 (with derived algebra of type A_1, corresponding to a short root) and nilradical of dimension 5.

As before, it is easy to locate the centralizer of x_β: it has a Levi factor of dimension 3 and nilradical of dimension 5. Thus the G-orbit of x_β has dimension 6; it is a new orbit, and not a Richardson orbit. By evaluating α and β at the conjugate $h_{3\alpha+2\beta}$ of h_β, we get respective values 0 and 1, leading to the weighted Dynkin diagram for the orbit named A_1 in the table.

orbit name	Dynkin weights	orbit dim	$\dim \mathfrak{l}$	$\dim \mathfrak{v}$	$\dim \mathfrak{c} \cap \mathfrak{l}$	$\dim \mathfrak{c} \cap \mathfrak{v}$
1	0 0	0	14	0	14	0
A_1	0 1	6	4	5	3	5
$\widetilde{\mathrm{A}}_1$	1 0	8	4	5	3	3
$\mathrm{G}_2(a_1)$	0 2	10	4	5	0	4
G_2	2 2	12	2	6	0	2

TABLE. Nilpotent orbits of G_2

Only one serious obstacle remains: How can we verify that there are no further nilpotent orbits? This clearly requires a more systematic approach, provided originally by Dynkin and later refined by Bala and Carter.

7.13. Distinguished Orbits and Parabolics

It is reasonable to look for an inductive approach to the enumeration of nilpotent orbits (or unipotent classes), relative to subsystems of the root system. Dynkin concentrated on the way in which nilpotent orbits intersect "regular" reductive subalgebras of $\mathfrak{g}$: reductive subalgebras containing a Cartan subalgebra (the Lie algebra of a maximal torus of G). If an orbit meets such a reductive algebra, it must meet the semisimple derived algebra—about whose nilpotent orbits we already have information (by induction on dimension) unless this is $\mathfrak{g}$ itself. But the remaining case of nilpotent orbits meeting no proper regular subalgebra is complicated to resolve.

While the notion of parabolic subalgebra was implicit in the classical theory of semisimple Lie algebras, it was not until the development of the Borel–Chevalley theory of semisimple algebraic groups that parabolic subgroups became central to the theory. Bala–Carter [**5, 6**] shift the focus from arbitrary regular reductive subalgebras to the smaller class of Levi subalgebras of parabolics. Given a nilpotent orbit, they look for the smallest possible Levi subalgebra which it intersects—which fortunately turns out to be well defined up to conjugacy:

PROPOSITION. *The Levi subalgebras of minimal dimension which meet a given nilpotent orbit in* $\mathfrak{g}$ *are all* G*-conjugate.*

As indicated above, one may hope to use induction on dimension to pin down those orbits which meet proper Levi subalgebras. But how are the remaining orbits to be detected?

To say that a nilpotent element fails to lie in any proper Levi subalgebra amounts to saying that its centralizer contains no nonzero semisimple element. We say that such an element (or its orbit) is **distinguished**. For example, regular nilpotent elements are distinguished (cf. 4.7). In the case of unipotent elements, a slight modification is needed because of the possible presence of a nontrivial center in G. We say that a unipotent element (or class) is distinguished if the centralizer contains no noncentral semisimple elements.

At first sight there is no obvious way to locate the distinguished orbits. From $\mathfrak{sl}_2$ theory one can see easily that the grading associated to such an orbit is characterized by the special property: $\dim \mathfrak{g}_0 = \dim \mathfrak{g}_2$. But a more subtle fact is needed: *all distinguished orbits are even.* The original proof of this by Bala–Carter was quite complicated, but Jantzen found a clever short proof which takes advantage of the existence of the Richardson dense orbit in the canonical parabolic of a given distingished element. (In turn, Proposition 7.9 above forces each distinguished orbit—being even—to be a Richardson orbit relative to the corresponding class of canonical parabolics.) This shifts the focus to the problem of characterizing the canonical parabolics (called **distinguished**) which belong to distinguished orbits. Here there is a clearcut criterion:

THEOREM. *Assume that* $\operatorname{char} K = 0$ *or* $p > 3(h-1)$*, where* h *is the Coxeter number. A parabolic subalgebra of* $\mathfrak{g}$*, with Levi decomposition* $\mathfrak{p} = \mathfrak{l} \oplus \mathfrak{v}$*, is distinguished if and only if* $\dim \mathfrak{l} = \dim \mathfrak{v}/[\mathfrak{v}, \mathfrak{v}]$*. Moreover, there is a natural one-to-one correspondence between distinguished nilpotent orbits and conjugacy classes of distinguished parabolic subalgebras in* $\mathfrak{g}$*.*

For example, Borel subalgebras are distinguished: writing $\mathfrak{b} = \mathfrak{t} \oplus \mathfrak{u}$, we have $\dim \mathfrak{t} = r = \dim \mathfrak{u}/[\mathfrak{u}, \mathfrak{u}]$. The corresponding distinguished orbit is the regular orbit.

This characterization makes it possible (with sufficient labor) to check from the root data whether each type of parabolic subalgebra is or is not distinguished. See 7.15 below for more discussion of low rank examples, and Carter [**29**, 5.9] for detailed tables. For the respective exceptional types G_2, F_4, E_6, E_7, E_8, the number of distinguished nilpotent orbits is $2, 4, 3, 6, 11$ (whereas the total number of orbits turns out to be $5, 16, 21, 45, 70$).

Remark. As in [**180**], Bala–Carter [**6**] work with the slightly more stringent bound $p \geq 4m + 3$, where $m = h - 1$ is the height of the highest root. The reworking of the classical Dynkin–Kostant theory in Carter [**29**, Chap. 5] permits an improvement of the bound to $3(h-1)$. But in any case, it takes further *ad hoc* work (see 7.16 below) to prove the main results for all good p.

7.14. Bala–Carter Theorem

The focus on distinguished orbits and parabolics is justified by the main result of Bala–Carter [**6**]:

THEOREM. *Assume that* $\operatorname{char} K = 0$ *or* $p > 3(h-1)$, *where h is the Coxeter number. Then there is a bijection between nilpotent orbits in $\mathfrak{g}$ and conjugacy classes of pairs* $(\mathfrak{l}, \mathfrak{q})$, *where $\mathfrak{l}$ is a Levi subalgebra of some parabolic subalgebra in $\mathfrak{g}$ and $\mathfrak{q}$ is a distinguished parabolic subalgebra in the derived algebra of $\mathfrak{l}$. In this bijection, the orbit associated to* $(\mathfrak{l}, \mathfrak{q})$ *is the one meeting $\mathfrak{q}$ in its Richardson orbit.*

Thanks to Theorem 6.20, a parallel theorem holds for unipotent classes in G. The proof of these theorems rests on the classical Lie algebra theory together with careful analysis of parabolic subalgebras and root systems.

Bala–Carter systematically label orbits (or classes) by symbols like those in the first column of the table in 7.12 above. The Lie type X_r identifies the type of a minimal Levi subalgebra $\mathfrak{l}$ meeting the orbit (with a tilde to distinguish a short root type if necessary), and the remainder of the label keeps track of the distinguished parabolic $\mathfrak{q}$ (other than the Borel subalgebra) within $\mathfrak{l}$: for example, a_1 just indicates that $\mathfrak{q}$ has semisimple rank 1. With some assistance from the tables of Elkington [**51**] (as corrected by Elašvili [**50**]), they correlate their classification with Dynkin's weighted diagrams. For the exceptional types, they provide full tables (reproduced with a couple of minor corrections in Carter [**29**, 13.1]).

Remark. As a byproduct of their approach, Bala–Carter [**6**, Thm. 7.1] are able to carry over to G a result on nilpotent orbit representatives in $\mathfrak{g}$ (see Springer–Steinberg [**180**, III, 4.29]): each unipotent element of G is conjugate to one of the form $\prod_\alpha x_\alpha(1)$, where the product is taken over a set of linearly independent positive roots. Moreover, a change in the order of the factors just yields another conjugate element. The possibility of choosing such class representatives over the prime field of K provides a good starting point for the study of unipotent classes in finite groups of Lie type, as will be seen in the next chapter. Even though the methods of Bala–Carter require $\operatorname{char} K = p > 3(h-1)$, one can still try out the same elements over fields of smaller (good) characteristic as plausible candidates for class representatives.

7.15. Examples Revisited

As one might expect, the Bala–Carter Theorem has a straightforward interpretation in the case of $\mathfrak{sl}(n, K)$ (for any K). The key observation is that only the Borel subalgebras and corresponding regular orbits in $\mathfrak{g}$ are distinguished: Take a nilpotent Jordan matrix which is not regular, so some $(k, k+1)$-entry is 0 rather than 1. It is easy to construct a nonzero diagonal matrix of trace 0 which commutes with this one: let the first k entries be equal to 1 and the remaining $n-k$ entries be equal to $-k/(n-k)$. (If char K is a prime dividing $n-k$, let the first k entries be 0, followed by $n-k$ entries equal to 1.)

On the other hand, each proper Levi subalgebra $\mathfrak{l}$ is conjugate to the algebra of block-diagonal matrices, with block sizes $\lambda_1 \geq \lambda_2 \geq \cdots \geq \lambda_d$ ($\sum \lambda_i = n$). Its derived algebra is therefore isomorphic to $\mathfrak{sl}(\lambda_1, K) \oplus \cdots \oplus \mathfrak{sl}(\lambda_d, K)$. Here the only distinguished parabolic is a Borel subalgebra, say the direct sum of upper triangular Borel subalgebras in the various blocks. Take a regular nilpotent element in each: a single $\lambda_i \times \lambda_i$ Jordan block. Thus we have associated to $\mathfrak{l}$ and its Borel subalgebra the nilpotent orbit corresponding to the partition λ. In this way the Bala–Carter Theorem (valid here in all characteristics) reproduces the usual parametrization.

The situation is more complicated for other classical or exceptional types: see Carter [**29**, 5.9] and Collingwood–McGovern [**31**, Chap. 8]. Here we just take a further look at the case of G_2. In 7.12 we were able to locate five nilpotent orbits (or unipotent classes) by a combination of methods. But we had no way to insure the completeness of our list. To apply the Bala–Carter method, we must first determine the distinguished parabolics. Besides the Borel subalgebra, we find that only the minimal parabolic corresponding to the short root α is distinguished. Here $\mathfrak{v}$ involves the positive roots other than α, and $[\mathfrak{v}, \mathfrak{v}] = [\mathfrak{g}_\beta, \mathfrak{g}_{3\alpha+\beta}] = \mathfrak{g}_{3\alpha+2\beta}$, yielding $\dim \mathfrak{v}/[\mathfrak{v}, \mathfrak{v}] = 4 = \dim \mathfrak{l}$. This yields the distinguished orbit (the subregular orbit) named $G_2(a_1)$ in 7.12.

Now it is easy to deal with proper Levi subalgebras: the derived algebra is either 0 (yielding the 0 orbit) or else a copy of $\mathfrak{sl}(2, K)$ corresponding to a long or short root (these Levi subalgebras being nonconjugate). In these two cases we recover the respective orbits named A_1 and $\widetilde{A_1}$.

7.16. Pommerening's Theorem

In two papers, Pommerening [**134, 135**] fills in the necessary details (using case-by-case arguments) to achieve an optimal version of the Bala–Carter Theorem 7.14:

THEOREM. *If* char K *is good for* G, *then the Bala–Carter Theorem is valid, giving a bijection between nilpotent orbits in* $\mathfrak{g}$ *and conjugacy classes of pairs*

$(\mathfrak{l}, \mathfrak{q})$, *where* $\mathfrak{l}$ *is a Levi subalgebra of some parabolic subalgebra in* $\mathfrak{g}$ *and* $\mathfrak{q}$ *is a distinguished parabolic subalgebra in the derived algebra of* $\mathfrak{l}$.

While the classical types can be handled fairly directly, the exceptional types require more intricate arguments. The main problem is to associate to each distinguished nilpotent element $x \in \mathfrak{g}$ a distinguished parabolic relative to which x is a Richardson element. Pommerening is able to embed x into a standard triple, but the representation theory of $\mathfrak{sl}(2, K)$ is not helpful for small primes. Instead he studies associated gradings of $\mathfrak{g}$, adapting methods of Vinberg and others from characteristic 0 and making many case distinctions.

It remains to be seen whether a unified conceptual proof of the Bala–Carter Theorem can be found, perhaps in the group (rather than Lie algebra) setting. In place of the heavy reliance on $\mathfrak{sl}_2$ theory, one might for example try to make better use of the flag variety.

In case p is a bad prime for G, case-by-case study (by many people) has led to a complete determination of unipotent classes in G as well as nilpotent orbits in $\mathfrak{g}$: see Carter [**29**, 5.11] for a careful summary with references to the literature. The number of such classes or orbits sometimes differs from the "generic" case, and sometimes differs between G and $\mathfrak{g}$.

7.17. Component Groups of Centralizers

Next we take a closer look at centralizers. As explained in 7.8, the weighted Dynkin diagram of a nilpotent element $x \in \mathfrak{g}$ encodes some basic information about $\mathfrak{c}_{\mathfrak{g}}(x)$. In particular, we can compute the dimensions of the nilradical and of a Levi subalgebra, each obtained by intersecting the centralizer with the canonical parabolic subalgebra. Further work is required to pin down the Lie type of the semisimple derived algebra of the Levi part. As long as the characteristic is good, there is a closely parallel description of the groups $C_G(x)$ or the corresponding (isomorphic) centralizers $C_G(u)$ of unipotent elements. But an important extra ingredient enters into this description: the component group $A(x) := C_G(x)/C_G(x)^\circ$ or $A(u) := C_G(u)/C_G(u)^\circ$. This turns out to play an essential role in work of Springer and Lusztig on representations of Weyl groups and finite groups of Lie type.

Here is a brief survey of what is known about the centralizers, from various angles. For classical types, direct descriptions of the centralizers and their Levi parts can in fact be given, independent of the Dynkin theory: see Collingwood–McGovern [**31**, 5.1.18, 6.1.3] for full details, following Wall [**198**], as well as Springer–Steinberg [**180**, Part IV]. In type A, all centralizers in question are connected, while in types B–D they sometimes have component groups which are elementary abelian 2-groups (direct products of groups of order 2). This is part of the calculation of the fundamental group of a nilpotent orbit in [**31**, 6.1].

As always, the exceptional types are more elusive. With these in mind, Elkington [**51**] developed the classical theory further and arrived at tables giving the

Levi parts of the centralizers. A number of errors in these tables were detected and corrected by Elašvili [**50**], who arrived at the results in a different way. Based on this work and Dynkin's original work, Alekseevskii [**2**] was then able to pin down the component groups in all exceptional cases. At about the same time Mizuno [**131**] was completing work on the type E cases (using a Chevalley group approach). Fortunately, these independently obtained results agree. The tables in Carter [**29**, 13.1] provide a detailed summary of all this information.

We shall have more to say about Mizuno's work in the next chapter, but it is appropriate to outline the methods of Alekseevskii here: they rely mainly on the classical work just surveyed. As in earlier sections, it has to be assumed throughout that $\operatorname{char} K$ is 0 or a sufficiently large prime.

The starting point is the observation that the component group $A(u) = C_G(u)/C_G(u)^\circ$ is the same as the component group L/L° of a Levi subgroup L of $C_G(u)$, while $L = C_G(S)$ for a subgroup S of type A_1 in which u is embedded. Moreover, the group L° is known (from the work of Elkington and Elašvili), together with data about its inner and outer automorphisms. The problem is to locate possibly disconnected subgroups of G which centralize S.

From Dynkin's work, supplemented by Lie group methods of de Siebenthal, one knows up to conjugacy the minimal "regular" reductive subgroups containing S. Such a group M is of maximal rank in G, and S lies in its semisimple derived group. Obviously the center Z of M centralizes S. It turns out that the images of such groups Z in L/L° are enough to generate the component group. To get more precise information, one has to analyze carefully several possible situations. There is a natural map from $A(u)$ to the outer automorphism group of L°, whose kernel is shown to be either trivial or else $A(u)$. The latter case occurs, for example, when all groups Z have the same dimension, and then the quotients Z/Z° are (up to conjugacy) all the maximal commutative subgroups of $A(u)$. The arguments here involve some delicate work on diagonalizable groups due to Steinberg [**189**].

Remark. After the fact one observes that all nontrivial component groups occurring for types A–G are elementary abelian 2-groups or else symmetric groups S_3, S_4, S_5 (with S_5 occurring only for a single class in E_8). The fact that the representation theory of these finite groups is especially simple has strong implications for Springer's representations of Weyl groups—for example, allowing the latter to be realized over $\mathbb{Q}$. See Chapter 9 below.

7.18. Another Look at G_2

One has to study a complicated group like E_8 to appreciate fully what is involved in carrying out the program of Alekseevskii. But even the group of type G_2 illustrates some of the typical phenomena. Here is another version of the table in 7.12 above (see Carter [**29**, 13.1]). The fourth column gives the Lie type of the Levi part of a centralizer (or 1 if this is trivial), while the last column gives the component group. This turns out to be trivial in all cases except the subregular class, where it is the symmetric group S_3.

class name	Dynkin weights	class dim	type of $C_G(u)$	$A(u) = C_G(u)/C_G(u)^\circ$
1	0 0	0	G_2	1
A_1	0 1	6	A_1	1
$\widetilde{\mathrm{A}_1}$	1 0	8	A_1	1
$\mathrm{G}_2(a_1)$	0 2	10	1	S_3
G_2	2 2	12	1	1

TABLE. Unipotent classes of G_2

Where does the S_3 comes from? From the viewpoint of [**2**], one looks for the minimal regular reductive subgroups containing a group S of type A_1 in which our unipotent element is embedded. Two non-conjugate ones occur in Dynkin's classification, and are in fact semisimple subgroups of types A_2 (generated by long root subgroups) and $\mathrm{A}_1 \times \widetilde{\mathrm{A}_1}$ respectively. (As in the table, the tilde indicates a short root type: here one has a subsystem of Φ involving two orthogonal roots such as β and $2\alpha + \beta$, one long and one short.) The respective centers of these two regular subgroups are found to be of orders 3 and 2. Alekseevskii's analysis then leads to the conclusion that these cyclic groups are maximal commutative subgroups of the component group, leading to S_3.

This case also illustrates a subtle question which came up earlier, concerning the relationship of centralizers of subregular elements to parabolics. Choose a subregular unipotent element u lying in some fixed Borel subgroup B, which in turn lies in minimal parabolic subgroups P_α (where α is the short simple root) and P_β (where β is the long simple root). Write $C := C_G(u)$. From 5.9 we know that C° is solvable, so C has S_3 as a Levi factor. (In fact, C° is unipotent.) We saw in 7.12 that P_α is the *canonical* parabolic associated with u, and therefore $C \subset P_\alpha$. But we also know that the subregular elements are Richardson elements in all minimal parabolics to which they belong (5.7). This implies that $C^\circ \subset P_\beta$ (Theorem 5.3). But not all of C is included in P_β. Indeed, part (iii) of Lemma 6.12 implies that $C \cap P_\beta$ has index 3 in C, since C permutes transitively three lines of type β in $(G/B)_u$. (This follows equally well from the direct computation of C in Elkington [**51**, p. 162].)

7.19. Closures of Orbits

It is a basic fact (1.7) that conjugacy classes in G are closed precisely when their elements are semisimple. At the other extreme, we know that the regular unipotent class is dense in the unipotent variety $\mathcal{U}$. There is a natural partial ordering on the collection of unipotent classes, defined as follows. If $u, v \in \mathcal{U}$, write $\mathrm{Cl}(u) \geq \mathrm{Cl}(v)$ if v lies in the closure of $\mathrm{Cl}(u)$. (The definition for nilpotent orbits is analogous.) We may call this the **closure ordering**. When char K is good, it is clear that the bijection of 6.20 between unipotent classes and nilpotent orbits is order-preserving.

As observed in 7.10 and 7.11, nilpotent orbits in Lie algebras of types A–D may be conveniently parametrized by partitions (as well as by weighted Dynkin diagrams). We again write $\lambda = [\lambda_1, \lambda_2, \ldots, \lambda_d]$ for a partition of n, with $\lambda_1 \geq \lambda_2 \geq \cdots \geq \lambda_d > 0$ and $\sum \lambda_i = n$. The partitions of n may be partially ordered as follows, using the convention that $\lambda_i = 0$ when i exceeds the number of parts of λ. Write $\lambda \geq \mu$ if for all k,

$$\sum_{i \leq k} \lambda_i \geq \sum_{i \leq k} \mu_i.$$

It is easy to check transitivity. When $n = 6$, for example, we have $[4,2] > [3,3] > [3,2,1]$ whereas $[4,1,1]$ and $[3,3]$ fail to be comparable. This partial ordering, often called the **dominance ordering**, arises in many combinatorial questions.

In particular, when $\mathfrak{g} = \mathfrak{sl}(n, K)$ the dominance ordering for partitions of n captures precisely the closure ordering for nilpotent orbits. The same turns out to be true for other classical types. (But some fine tuning is required in the case of D_r for r even, where certain partitions actually parametrize two distinct orbits: these are of the same dimension, hence are not comparable in the closure ordering.) The following theorem is essentially due to Gerstenhaber [**59, 60**], whose arguments were not quite complete and were later reworked by Hesselink [**68**, 3.10].

THEOREM. *Let $\mathfrak{g}$ be of classical type, with nilpotent orbits $\mathcal{O}_\lambda$ parametrized by suitable partitions. Then $\mathcal{O}_\lambda > \mathcal{O}_\mu$ if and only if $\lambda > \mu$.*

It is easy to sketch the proof for $\mathfrak{sl}(n, K)$. The first step is combinatorial, involving the ranks of powers of a nilpotent matrix (which are constant on its orbit). Say x_λ is the nilpotent Jordan matrix with Jordan blocks of sizes $\lambda_1, \lambda_2, \ldots, \lambda_d$. Observe that x_λ^k has rank equal to $\sum(\lambda_i - k)$, where the sum is taken over those i for which $\lambda_i \geq k$. Then it follows from the definition of the dominance ordering that

$$\lambda \geq \mu \text{ if and only if } \operatorname{rank} x_\lambda^k \geq \operatorname{rank} x_\mu^k \text{ for all } k.$$

The next step is more geometric. Recall that the rank of a matrix is the size of the largest possible nonzero minor. Moreover, the vanishing of all minors of a given size is equivalent to the vanishing of certain polynomials in the matrix entries (determinants). It follows that if x_μ lies in the closure of the orbit of x_λ, then we must have

$$\operatorname{rank} x_\lambda^k \geq \operatorname{rank} x_\mu^k \text{ for all } k.$$

More careful study, using $\mathfrak{sl}_2$ theory for the associated standard triples, shows that the converse holds as well. Combined with the first step, this characterizes the closure ordering in terms of the dominance ordering of partitions.

(Collingwood–McGovern [**31**, 6.2] give full details for type A, but refer to Hesselink for other cases. See also Carter [**29**, 13.1].)

The ideas involved here are uniform whenever char K is good. Even though the results for classical types B–D in characteristic 2 are slightly different, Spaltenstein [**160**, I.2.10, II.8.2] extends the Gerstenhaber–Hesselink methods to this situation as well.

We remark that a different—possibly more intuitive—approach for $\mathfrak{sl}(n, K)$ might be contemplated. Consider again a nilpotent Jordan matrix x_λ, with an entry 1 in the $(k, k+1)$ position. The adjoint action of a suitable diagonal matrix t in $\mathrm{SL}(n, K)$ will change this position to any prescribed nonzero scalar c while leaving other entries intact. Indeed, we can choose an nth root of c^{-k} in K (call it b), set $a = bc$, and define $t := \mathrm{diag}(a, \dots, a, b, \dots, b)$, with k occurrences of a. It is easy to check that $\det t = 1$ and that $\mathrm{Ad}\, t$ acts correctly. Thus the orbit of x_λ contains a sort of affine line with 0 removed, and the closure of the orbit must contain the Jordan matrix obtained by replacing 1 with 0 in the $(k, k+1)$ entry. Its partition μ (re-ordered if necessary so the parts are in descending order) then satisfies $\lambda > \mu$. For example, when $n = 6$ and $\lambda = [3, 3]$, take $k = 5$ to produce $\mu = [3, 2, 1]$. Another example is the regular orbit: from $\lambda = [n]$ one can obtain any other partition by a sequence of the steps just described.

As usual, the exceptional types E–G require separate treatment. In all five cases, the closure ordering has been worked out in the group setting (mainly by Mizuno [**131**] in the complicated cases E_6, E_7, E_8). His diagrams are reproduced in Spaltenstein [**160**, Chap. IV] as well as Carter [**29**, 13.4]. For G_2, the ordering is linear (as given in the table in 7.12 above). But the other types lead to increasingly intricate diagrams. Fortunately, some independent later work of Beynon and Spaltenstein [**11**] described in [**160**, Addendum] confirms Mizuno's results.

One last comment: There is a certain amount of hidden symmetry in these order diagrams, which was discovered by Lusztig in a very different context (see Spaltenstein [**160**, Chap. III], as well as Carter [**29**, 13.4]). Certain orbits are regarded as "special", and when all others are discarded the order diagram has a top–bottom symmetry revealing a type of duality. Though quite subtle in the exceptional cases, this duality takes on a very classical form for type A: here all orbits are special, and the duality is achieved simply by taking the conjugate of a partition.

7.20. Geometry of Orbit Closures

Classes in G and orbits in $\mathfrak{g}$ are smooth locally closed subsets, in most cases isomorphic to quotient varieties $G/C_G(x)$. But their closures may acquire singularities. This is seen first in $\mathrm{SL}(2, K)$, where the closure of the regular class is the unipotent variety—a cone with a singularity at e. The geometry of class or orbit closures has been extensively studied, especially in characteristic 0. This is of course part of a broader study of orbits occurring in G-actions on affine varieties, in particular actions arising from linear representations of G (see Borho–Kraft [**20**]).

Since semisimple classes are closed, attention focuses mainly on unipotent classes (or nilpotent orbits). If the closure of a class or orbit fails to be smooth, one may ask next whether it is *normal*, and if so whether its singularities are rational. (In terms of sheaf theory, a normal variety Z is said to have **rational singularities** if there is a desingularization $\varphi : Y \to Z$ such that $R^i\varphi_*\mathcal{O}_Y = 0$ for all $i > 0$.) These are subtle questions, usually requiring case-by-case analysis.

The most definitive results have been obtained when $\operatorname{char} K = 0$, starting with the fundamental paper of Kostant [**98**] and continuing with the work of Borho, Kraft, Procesi, and others. In §2 of his paper Kostant developed a representation-theoretic framework for the study of normality of orbit closures (see also Borho–Kraft [**20**, 3.7] and Kraft [**101**, §5]). Briefly put, the algebra of regular functions on an orbit is the integral closure of the algebra of regular functions on its closure—provided all orbits in the boundary are of codimension at least 2. So these algebras coincide just when the orbit closure is normal. Since an orbit is isomorphic to the quotient of G by an isotropy subgroup H, the regular functions on the orbit are the H-invariants in $K[G]$ and may be studied using representation theory. In his paper Kostant proved that the nilpotent variety is normal, and is the closure of a single orbit: the regular nilpotent orbit.

As late as 1977 it seemed possible that all orbit closures might be normal. Hesselink [**71**] (see also his related papers [**68, 69, 70**]) worked out some low rank cases in which this is true. In a series of papers, Kraft and Procesi [**100, 102, 103, 104**] studied the classical groups thoroughly. In the case of $\mathrm{GL}(n, K)$ they showed that all orbit closures are normal, with rational singularities. They also described precisely the "minimal" singularities (those occurring in open classes in the boundary of an orbit closure), using a mixture of algebraic geometry and classical invariant theory. But for groups of types B, C, D, not all orbit closures are normal: precisely which are and are not was worked out in [**104**].

Kraft [**101**] completed the picture in the case of groups of type G_2 (still working in characteristic 0): only one of the five nilpotent orbits (denoted $\widetilde{\mathrm{A}}_1$ in 7.12) fails to be normal. This paper also contains a useful summary (in §8) of results on the normality problem. As he points out, calculations of Beynon and Spaltenstein [**11**] determine those nilpotent orbits in groups of type E_r having a branched non-normal closure. Further work of Richardson [**139**] for type E_r detected non-normal closures for which the normalization map is bijective.

Much less seems to be known when $\operatorname{char} K = p > 0$. In the case of $\mathrm{GL}(n, K)$, Donkin [**48**] was able to modify the approach of Kraft–Procesi [**102**], thereby proving the normality of all conjugacy class closures. In characteristic 0 the complete reducibility of finite-dimensional rational representations of G plays a key role in determining how G acts on rings of regular functions. This breaks down badly in prime characteristic. But a weaker property of locally finite-dimensional modules turns out to be adequate: having a "good filtration" (with filtration quotients isomorphic to modules induced from a Borel subgroup).

Mehta and van der Kallen [**128, 129**] have gone somewhat further in this case, by using instead the notion of "Frobenius splitting" for varieties in characteristic p. The desingularization of the unipotent variety (or nilpotent variety) studied in Chapter 6 provides a normal variety mapping onto the closure of an orbit. In the case of $\mathrm{GL}(n, K)$, they showed that this map admits a Frobenius splitting, which in turn implies that the orbit closure is normal and has rational singularities.

CHAPTER 8

Finite Groups of Lie Type

In this chapter $\operatorname{char} K = p > 0$ and k is a finite subfield of order $q = p^n$, over which the connected semisimple group G is defined.

The goal here is to study the behavior of classes and centralizers in G relative to k, with applications to the finite group $G(k)$. (In practice, we will soon denote the finite group by G^F, following the widespread practice of describing groups of Lie type as fixed points of Frobenius maps.) While a few of the results in this chapter might be proved entirely within the framework of finite group theory, the reference to G almost always illuminates what is going on.

The main facts about classes and centralizers will be described in some detail, with references to the literature for full proofs. After a certain point lengthy case-by-case analysis is needed, for example to count the total number of classes. We limit ourselves to giving a careful survey of methods and results, avoiding too much overlap with the books of Carter [**29**] and Digne–Michel [**46**]. Besides these books, our major sources include Steinberg [**187**], Springer–Steinberg [**180**].

It becomes clear very soon that the study of classes in $G(k)$ interacts frequently and sometimes in unexpected ways with the study of representations of $G(k)$—both ordinary and modular. The total number of classes is, for example, well-known to be the same as the number of irreducible characters. Similarly, the number of classes of semisimple elements is equal to the number of inequivalent irreducible representations of $G(k)$ over K. Less obviously, properties of the Steinberg character lead to a simple computation of the number of unipotent elements in $G(k)$.

8.1. Some Natural Questions

Whenever a semisimple (or reductive) group G is defined over a subfield k of an algebraically closed field K, a number of questions arise immediately about the relationship between classes of G and of $G(k)$. Here we are interested in the case of a finite field k of order q, though many of the same issues would have to be dealt with generally. A special feature of the finite case is the possibility of counting things—the group order being one example. Often the result is a polynomial in q.

Roughly speaking, one wants to know first which classes (or centralizers) are defined over k, and then how many rational points they have. Since k is perfect, these questions are simplified somewhat: being k-closed is equivalent is being defined over k. For example, the centralizer in G of any element of $G(k)$ is automatically defined over k.

Here are some specific questions to be addressed in the finite case:

- Which classes in G are defined over k?
- If a class in G is defined over k, must it contain an element of $G(k)$?
- If a class C in G intersects $G(k)$, how does $C \cap G(k)$ split into classes in $G(k)$?
- What is the size of each class (or centralizer) in $G(k)$?
- How many semisimple (resp. unipotent) classes does $G(k)$ have?
- What is the total number of classes in $G(k)$?

Through the efforts of many people, most of these questions have been answered in satisfactory detail. The answers often require case-by-case proofs for the simple types, and may depend critically on the isogeny type of G, or whether p is a *good* prime (3.9), or whether q satisfies one or another congruence. Sometimes the methods are rather messy and *ad hoc*, with details omitted or proofs based on computer calculations. Moreover, there are occasional errors or misprints in published results (such as tables involving unipotent classes). While enough cross-checking has been done to permit a high level of confidence in the central results—or at least a convergence toward a consensus—the reader should approach the literature with a reasonable degree of caution.

As in previous chapters, the Jordan decomposition plays a natural organizing role. This coincides in $G(k)$ with the type of decomposition found in any finite group: semisimple elements are those of order not divisible by p (called **p-regular elements**), unipotent elements are those of order equal to a power of p (called **p-elements**), and every element is uniquely a product of two commuting elements of these respective types.

8.2. Frobenius Maps and Finite Groups

From the perspective of algebraic groups, most finite groups of Lie type are close relatives of the groups $G(k)$ of rational points of simple algebraic groups defined over finite fields k. Lang's Theorem (8.3 below) implies that G must be at least **quasi-split** over k: it has a Borel subgroup defined over k. In the **split** case G has a k-split maximal torus T (diagonalizable over k), which implies that G is quasi-split. More precisely, any k-split maximal torus is contained in a k-split Borel subgroup B, whose root groups U_α are k-isomorphic to the additive group.

In the quasi-split (but not split) case G still has a Borel subgroup and maximal torus defined over k, but the behavior of the root system relative to k is more complicated. This involves a nontrivial symmetry of the Dynkin diagram—

possible only in types A_r (for $r > 1$), D_r, E_6. Symmetries of order 2 lead to new finite groups denoted by ^2A_r, ^2D_r, $^2E_6$, while a symmetry of order 3 leads to $^3D_4$.

Except in a few small cases, the resulting finite groups $G(k)$ yield simple groups when G is simple: By passing to the derived group and factoring out the center, one almost always gets a nonabelian simple group. In the case of split adjoint groups, Chevalley provided a direct uniform construction in his famous 1955 Tôhoku paper, starting with a Chevalley basis for the corresponding simple Lie algebra over $\mathbb{C}$. So the split groups are often known as **Chevalley groups**. Later work of Hertzig, Steinberg, Tits added the remaining quasi-split groups of types A, D, E to the list; these are often called **twisted groups**.

There are, however, some additional finite groups of Lie type in characteristics 2 and 3 associated with the root systems B_2 ($= C_2$), F_4, G_2. Discovered by Suzuki and Ree, these require a modified construction related to Chevalley's exceptional isogenies (cf. 0.13). Here there is an extra twist which interchanges the short and long roots. Though not so directly related to groups of the form $G(k)$, these groups still have many of the same structural features (notably a Tits system).

Comprehensive treatments of the finite groups from the Chevalley–Steinberg viewpoint—rather than the algebraic group viewpoint—are given by Steinberg [**186**] and Carter [**26**].

In his fundamental paper [**187**], Steinberg provides a more unified framework for studying the structure and representations of all these groups of Lie type. This starts with the observation in the split case that $G(k)$ may be regarded as the set of fixed points of the map which raises each entry of G to the qth power (in a fixed matrix realization over k). Of course, this has to be formulated in terms of the algebra of regular functions on G to be intrinsic. In the quasi-split case, one has to modify the map by combining it with an automorphism of G coming from the diagram symmetry. A further modification leads to the description of Suzuki and Ree groups as fixed point sets.

With this in mind, Steinberg classifies all those surjective endomorphisms $\sigma : G \to G$ of the simple algebraic group G which lead to finite fixed point groups G_σ. These turn out to correspond precisely to the above families of finite groups of Lie type. One great advantage of working systematically with σ is that it leaves stable many subgroups of G, and induces related maps on objects such as the Weyl group and the character group of a maximal torus.

In recent literature the tendency has been to call such an endomorphism of G a **Frobenius map**, denoted F, and to write G^F for the group of fixed points. This is what we shall do throughout this chapter. The qth power map is sometimes referred to as a "standard" Frobenius map. For further details, including comparisons with standard labels for classical groups, see Carter [**29**, 1.17–1.19] and Digne–Michel [**46**] (who usually leave aside the Suzuki and Ree groups to simplify their presentation, as we shall also do).

8.3. Lang's Theorem

A theorem of Lang [**107**] (later generalized by Steinberg) plays a crucial role in the study of G^F and its conjugacy classes. In its original form, the theorem involves a connected algebraic group H defined over the field $k = \mathbb{F}_q$. If H is viewed as a matrix group, with standard Frobenius map $F : H \to H$ raising all matrix entries to the qth power, then the conclusion is that the morphism $\lambda : H \to H$ defined by $\lambda(x) = x^{-1}F(x)$ is *surjective.* (This is sometimes written with x replaced by x^{-1}, which doesn't affect the conclusion.)

In order to accommodate Frobenius maps in the more general sense defined above, Steinberg [**187**, §10] reformulates Lang's statement as follows:

THEOREM. *Let H be a connected algebraic group, and let $F : H \to H$ be a surjective endomorphism fixing only finitely many elements of H. Then the morphism $\lambda : H \to H$ sending x to $x^{-1}F(x)$ is surjective.*

A couple of remarks are in order:

(1) As Steinberg makes explicit in [**184**], the theorem is in effect a vanishing theorem for Galois cohomology. There a general vanishing theorem of this sort is proved for connected algebraic groups over a field of "cohomological dimension" 1 (finite fields being one example).

(2) Notice that the finiteness of the fixed point group of F is a necessary as well as sufficient condition for the surjectivity of λ: surjectivity of F insures that fibres are generically finite. In investigating the possibilities for surjective endomorphisms F when H is simple, Steinberg concluded that either the fixed point group is finite or else F is an automorphism.

Although the statement of Lang's Theorem is not intuitively obvious, the idea of the proof is fairly straightforward. The crucial observation is that the differential of F is zero when F is the standard Frobenius map, and is in any case nilpotent under the more general assumptions here. Now the differential of λ at e sends $z \in \mathfrak{g}$ to $-z + (dF)_e(z)$, so the nilpotence of dF implies the surjectivity of $d\lambda_e$. In turn, λ is dominant and separable, so its image contains a nonempty open subset of H. By similar reasoning, if $h \in H$ is any fixed element, the morphism $x \mapsto x^{-1}hF(x)$ has a surjective differential and thus its image also contains a nonempty open subset of H. Because H is connected, these open sets have an element in common:

$$x^{-1}F(x) = y^{-1}hF(y),$$

forcing $h = z^{-1}F(z)$ with $z := xy^{-1}$.

Later Steinberg [**191**] offered a different approach to the proof, bringing out the fact that λ is a *finite* morphism. For other accounts of the theorem and its consequences, see B§16, Springer–Steinberg [**180**, I, 2.2], Carter [**29**, 1.17], Digne–Michel [**46**, 3.10]. In the following two sections we discuss explicitly the consequences of immediate interest for G^F.

8.4. Fixed Points in F-Stable Classes

One important application of Lang's Theorem is the following:

THEOREM. *Every F-stable conjugacy class in G^F contains an element of G^F.*

The proof is easy. If C is an F-stable class, let $x \in C$ be arbitrary. By hypothesis, $F(x) \in C$, so there exists $g \in G$ for which $gF(x)g^{-1} = x$. Now use Lang's Theorem to write $g = h^{-1}F(h)$ for some $h \in G$. We assert that the conjugate hxh^{-1} of x lies in G^F. Indeed,

$$x = gF(x)g^{-1} = h^{-1}F(h)F(x)F(h)^{-1}h = h^{-1}F(hxh^{-1})h,$$

so F fixes hxh^{-1}.

This result is just a special case of a more general principle based on Lang's Theorem: when a connected algebraic group defined over k acts on a k-variety, then any orbit defined over k (stable under the standard Frobenius map) contains a rational point. One further consequence of this principle, in our setting, is the fact that when H is a closed connected subgroup of G stable under F, we have $(G/H)^F \cong G^F/H^F$. In a similar spirit, one proves that G has a Borel subgroup defined over k (all of these being G^F-conjugate), and that such a Borel subgroup in turn contains a maximal torus of G defined over k. All of this is discussed in more detail in §3 of Digne–Michel [**46**]. As stated above, Lang's Theorem is in effect a vanishing theorem in Galois cohomology, where a typical issue is whether homogeneous spaces for G defined over k must contain rational points.

8.5. Classes Meeting G^F

As just seen, Lang's Theorem is crucial for proving that each F-stable class C in G meets G^F. It is also the key tool needed to sort $C^F := C \cap G^F$ into G^F-classes. This requires more subtle arguments.

Whenever F acts on a group A (which might be a closed subgroup A of G, or a subquotient such as A/A°), there is an action of A on itself: $a \in A$ sends $b \mapsto abF(a)^{-1}$. This is of course just conjugation if F acts trivially on A, which is often true in cases of interest. Denote by $\mathrm{H}^1(F, A)$ the collection of orbits in A for this action, called F-**conjugacy classes**.

THEOREM. *Set $C := \mathrm{Cl}_G(x)$ and $Z := C_G(x)$. Write C^F/G^F for the collection of G^F-classes in C^F. Then there are bijections φ and ψ:*

$$C^F/G^F \xrightarrow{\varphi} \mathrm{H}^1(F, Z) \xrightarrow{\psi} \mathrm{H}^1(F, Z/Z^\circ).$$

In particular, if $C_G(x)$ is connected, C meets G^F in a single class.

Here is a brief outline of the steps involved. To define φ, take an element $y = gxg^{-1}$ $(g \in G)$ in one of the G^F-classes in C^F. Applying F, one has $y = F(g)xF(g^{-1})$, forcing $g^{-1}F(g) \in Z$. Then φ maps the G^F-class of y to the F-conjugacy class of $g^{-1}F(g)$. Routine checking shows that this recipe is independent of the choice of y in its G^F-class (since F fixes elements of G^F) and of the choice of g (since another element conjugating x to y would be congruent to g modulo Z, yielding the same F-conjugacy class).

It is also routine (but messy) to verify that φ is one-to-one. That φ is surjective follows from Lang's Theorem: each $z \in Z$ can be written as $z = g^{-1}F(g)$ for some $g \in G$, and one then checks that $gxg^{-1} \in C^F$, so the F-conjugacy class of z is the image under φ of the G^F-class of gxg^{-1}.

Obviously ψ is surjective. To show it is one-to-one, a more devious application of Lang's Theorem is needed. Suppose the classes of $a, b \in Z$ map to the same F-conjugacy class in Z/Z°. This means that $a \equiv bF(c)^{-1} \pmod{Z^\circ}$ for some $c \in Z$. We may replace b by $cbF(c)^{-1}$ and assume $a \equiv b \pmod{Z^\circ}$. By Lang's Theorem, $b = g^{-1}F(g)$ for some $g \in G$, and then F stabilizes the connected group $gZ^\circ g^{-1}$. Applying Lang's Theorem to this group, and using the fact that $ab^{-1} \in Z^\circ$, we get $c \in Z^\circ$ for which

$$g(ab^{-1})g^{-1} = (gcg^{-1})^{-1}F(gcg^{-1}).$$

Using $b = g^{-1}F(g)$, this yields $a = c^{-1}bF(c)$, and thus a and b are F-conjugate.

For more details, see Springer–Steinberg [**180**, I, 2.6–2.8, 3.4] and Digne–Michel [**46**, §3].

Taken together, Theorems 8.4 and 8.5 provide a reasonable approach to the determination of classes in G^F, assuming one can pick out the F-stable classes in G. An easy example is the regular unipotent class of G, discussed in 8.12 below. However, when a class in G intersects G^F in more than one class, it requires more work to determine the sizes of these classes.

8.6. An Easy Example: SL$(2, q)$

Although it is a long way from A_1 to E_8, even the rank 1 case illustrates many of the significant theorems to be discussed below. Take $G = \mathrm{SL}(2, K)$ and $q = p^n$. First, one sees easily that $|\,\mathrm{SL}(2, q)| = q(q-1)(q+1)$, a polynomial in q of degree $\dim G$.

If F is the standard Frobenius map (raising matrix entries to the qth power), there are two G^F-classes of F-stable maximal tori in G, corresponding to the two conjugacy classes of the Weyl group S_2. The split (diagonal) torus intersects $G^F = \mathrm{SL}(2, q)$ in a group of order $q - 1$, while a nonsplit torus (split over a field of order q^2) intersects G^F in a group of order $q + 1$. All semisimple elements in G^F are conjugate to elements of these two tori. Apart from the identity matrix and (if $p \neq 2$) its negative, which comprise $Z(G)$, the semisimple elements are all regular and therefore those lying in G^F have centralizers of order $q - 1$ or $q+1$. Since elements of a maximal torus are conjugate in G precisely when they are conjugate under W, we conclude that G^F has q semisimple classes in all (no matter what p is).

Apart from e, all unipotent elements in G are regular and are conjugate to elements of the upper triangular group U. The centralizer of such an element in

U is just $UZ(G)$. Obviously the regular unipotent class meets G^F, and therefore (by Theorem 8.5) determines a single class of size $q^2 - 1$ in G^F when $p = 2$ or two classes when p is odd. As class representatives in the latter case one can take

$$\begin{pmatrix} 1 & 1 \\ 0 & 1 \end{pmatrix} \text{ and } \begin{pmatrix} 1 & c \\ 0 & 1 \end{pmatrix},$$

where c is a nonsquare in k. Since the centralizer order in each case is $2q$, the class size is $(q^2 - 1)/2$. Taking account of e, we find a total of q^2 unipotent elements in G^F (for all p).

Finally, the total number of classes in G^F is readily seen to be $q + 4$ (when p is odd) or $q + 1$ (when $p = 2$). (For more details, see Humphreys [**83**].)

For the adjoint group $\mathrm{PGL}(2, K)$, the results are closely analogous but not precisely the same—unless $p = 2$, in which case $\mathrm{SL}(2, q)$ is isomorphic to $\mathrm{PGL}(2, q)$. The group orders always agree, as a result of a general theorem (see B16.8) for isogenous groups over finite fields: $|\,\mathrm{PGL}(2, q)| = q(q-1)(q+1)$. The total number of unipotent elements is again q^2, but for odd p we find just two unipotent classes. The number of F-stable semisimple classes is still q, as in the case of $\mathrm{SL}(2, K)$. But the number of semisimple classes and the total number of classes in $\mathrm{PGL}(2, q)$ vary slightly from the case of $\mathrm{SL}(2, q)$ when p is odd: there are $q+1$ semisimple classes, and $q + 2$ classes in all. The discrepancy in the number of semisimple classes reflects the fact that not all semisimple elements in $\mathrm{PGL}(2, K)$ have connected centralizers, allowing a class to split into two classes in $\mathrm{PGL}(2, q)$ (as illustrated in 1.12).

8.7. The Number of Semisimple Classes

Now we return to the general case, and see what can be said about the semisimple classes in G^F. The following fundamental theorem is due to Steinberg [**187**, 14.8].

THEOREM. *Let G be a connected semisimple group of rank r, defined over $k = \mathbb{F}_q$, with associated Frobenius map F.*

(i) *The number of F-stable semisimple classes in G is q^r.*

(ii) *If G is simply connected, then G^F has q^r semisimple conjugacy classes.*

Here are a few comments on the proof (for details, see Carter [**29**, 3.7] as well as [**187**] and Springer–Steinberg [**180**, II, §3]).

The semisimple classes in G are in one-to-one correspondence with W-orbits in a maximal torus T (Proposition 3.1). Taking T to be F-stable, we look for the number of W-orbits fixed by F. This leads to some combinatorics related to the natural reflection representation of W and its exterior powers, resulting finally in (i). (Lehrer [**108**, §2] has an alternative approach, giving precisely parallel results in the Lie algebra case.)

Assuming that G is simply connected, all centralizers of semisimple elements are *connected*, by Theorem 2.11. In turn, Theorems 8.4 and 8.5 (based on Lang's

Theorem) imply that F-stable semisimple classes are in bijection with semisimple classes of G^F, so (ii) follows from (i).

In the simply connected case, the theorem was in fact proved earlier and more directly by Steinberg in [**183**, §7, §9], as part of his study of representations of G and G^F in characteristic p. Recall from 3.5 that the Steinberg map $\pi : G \to \mathbb{A}^r$ induces a bijection between semisimple classes in G and points of $\mathbb{A}^r$. The map involves evaluation of fundamental characters (here G must be simply connected!), and is defined over k when G is. In this way the F-stable semisimple classes are in bijection with the q^r rational points of $\mathbb{A}^r$ over k.

The significance for the representation theory of G^F is this. By a well-known theorem of Brauer, for any finite group H and any prime p, the number of p-regular conjugacy classes in H (those consisting of elements of order prime to p) is the same as the number of inequivalent irreducible representations of H over a sufficiently large field of characteristic p. Motivated by this viewpoint, Curtis [**32**, 5.10] gives another proof of part (ii) of the theorem, avoiding algebraic groups. He works in the somewhat more general axiomatic framework of groups with a (B,N)-pair (or Tits system), applicable to G^F, where an appropriate notion of "highest weight" of a representation can be formulated intrinsically without the presence of an ambient algebraic group.

Approaching from another direction, Deriziotis [**37**] (see Carter [**29**, 3.8]) recovers part (ii) of the theorem as a byproduct of his study of centralizers: see 8.10 below.

8.8. F-Stable Maximal Tori

Parametrizing semisimple classes of G by points of $\mathbb{A}^r$ (when G is simply connected) makes it easy to count the classes defined over k, but does not immediately provide insight into the sizes of the classes. Over K there is a hierarchy of centralizers, ranging from a maximal torus to the full group G. The situation is more complicated over k: even the regular semisimple elements in G^F may lie in maximal tori which fail to be conjugate over k. To get some insight into the behavior of centralizers relative to k, we therefore begin with tori. Here G can be any connected semisimple (or reductive) group.

As a consequence of Steinberg's generalized version of Lang's Theorem 8.3, G possesses F-stable maximal tori, at least one of which lies in an F-stable Borel subgroup. Such a torus is called **maximally split**. It is k-split just when G is k-split (so that G^F is a Chevalley group).

Now the F-stable maximal tori are permuted by the conjugation action of G^F—but not transitively. To obtain a parametrization of the orbits, fix a maximally split F-stable maximal torus T, with normalizer N in G and corresponding Weyl group $W = N/T$. Then F acts naturally on both N and W. A little calculation shows that a conjugate ${}^gT$ is F-stable just when $g^{-1}F(g) \in N$, leading to an element $w \in W$. Moreover, the G^F-class of ${}^gT$ is determined precisely by the F-conjugacy class of w as defined in 8.5: here w is F-conjugate to w' if there is some $z \in W$ for which $w' = z^{-1}wF(z)$. In the reverse direction, each F-conjugacy class in W yields a well-defined G^F-class of F-stable maximal tori: by Lang's theorem, any representative $n \in N$ of $w \in W$ can be written as $n = g^{-1}F(g)$ for some $g \in G$, leading to an F-stable torus ${}^gT$. The torus ${}^gT$ is said to be obtained from T by "twisting" with w and may also be denoted by T_w. To summarize:

PROPOSITION. *The G^F-classes of F-stable maximal tori in G are in natural bijection with the F-conjugacy classes in W(which are just the classes of W when G is split over k). Here the maximally split tori correspond to the trivial class of W.*

For example, when $G = \mathrm{SL}(2, K)$, there are two classes in $W = S_2$, corresponding to the two classes of maximal k-tori defined over k: those which are diagonalizable over k and those which become diagonalizable only over a quadratic extension (cf. 8.6 above).

Having sorted out the F-stable maximal tori in this way, one can go on to compute the order of each T_w^F. The idea is to study the induced actions of F and w on the group $Y(T)$ of one-parameter subgroups, or the vector space obtained by extending scalars to $\mathbb{R}$. The end result is that $|T_w^F|$ is obtained (when G is split) by evaluating the characteristic polynomial of w at q. In the case of $\mathrm{SL}(2, q)$, the split and nonsplit tori therefore have $q-1$ and $q+1$ rational points, respectively. In twisted cases, a further modification is needed. For example, in $\mathrm{SU}(n, q)$ one replaces q by $-q$ in each formula. This is all explicitly computable from the root system. See for example Deriziotis–Fakiolas [**42**] for groups of type E_r.

For detailed accounts of the theory just summarized, see Steinberg [**187**, §14], Springer–Steinberg [**180**, II, §1], Carter [**29**, 3.3.3], Digne–Michel [**46**, Chap. 3].

An obvious further problem is to count the number of F-stable maximal tori. Steinberg proves [**187**, 14.14]:

THEOREM. *The number of F-stable maximal tori in G is $q^{|\Phi|}$.*

For an exposition of the proof, which uses heavily the invariant theory of W in its natural representation on polynomials, see Carter [**29**, 3.4]. A recent streamlined proof by Lehrer [**108**, 1.11] has the added advantage of showing in parallel fashion that the number of F-stable maximal toral subalgebras of $\mathfrak{g}$ is also $q^{|\Phi|}$.

8.9. Regular Semisimple Classes

The F-stable maximal tori in G are the connected centralizers of regular semisimple elements in G^F. To count the number of regular semisimple classes in G^F, one can therefore concentrate on a fixed F-class of tori. Fixing as above a maximally split torus T, let T_w be obtained from it by twisting by $w \in W$. The number of classes of regular semisimple elements in G^F having connected centralizer conjugate to T_w is quickly seen to be equal to the number of regular elements in T_w (denoted $|T(w,\emptyset)|$) divided by the order of the "F-centralizer" of w in W.

The number $|T(w,\emptyset)|$ turns out to be a monic polynomial in q of degree r. To compute these numbers explicitly for a given group G, one has to get heavily involved in the combinatorics of the root system and Weyl group: Möbius inversion for the poset of w-stable subsystems of Φ, as well as Poincaré polynomials associated with hyperplane complements for W.

Lehrer [**108**] develops in some generality the connections of this and similar counting problems in G^F (or $\mathfrak{g}^F$) to W and its class functions, as well as to the topology of hyperplane complements.

Fleischmann and Janiszczak [**55**] focus on classical types A–D: simply connected groups and standard realizations such as $\mathrm{SO}(n,q)$, as well as twisted groups arising from graph automorphisms of order 2. Using their earlier calculations of Möbius functions [**54**], they obtain explicit expressions for the number of regular semisimple classes having each possible type of twisted maximal torus as centralizer.

8.10. Centralizers of Semisimple Elements

The methods used to study regular semisimple classes can be generalized in a natural way for arbitrary semisimple classes. Again one starts with centralizers. The theory as developed by Steinberg in [**187**] already permits some explicit computations of centralizer orders for semisimple elements of G^F. Burgoyne and Williamson [**25**] take some concrete steps in this direction for Chevalley groups. But the refined analysis of Carter and Deriziotis leads to far more comprehensive calculations.

Recall from Chapter 2 what is known about centralizers of semisimple elements in G: If $s \in G$ is semisimple, $C_G(s)$ is a reductive subgroup generated by a maximal torus T together with root groups $U_{\pm\alpha}$ for which $\alpha(s) = 1$ and Weyl group representatives in $N(T)$ commuting with s (2.2). If G is simply connected, $C_G(s)$ is connected, and therefore generated already by T and various root groups $U_{\pm\alpha}$ (2.11).

To simplify the exposition, say G is *simply connected*, so the groups $C_G(s)$ are connected. In this case there are q^r semisimple classes in all (8.7), a number which depends on q and therefore can be arbitrarily large. But even if two semisimple elements of G^F fail to be conjugate, their centralizers may well be conjugate, as in the regular case. So we should again look for a classification in terms of G^F-orbits. To be precise, we say that two semisimple elements $s, t \in G^F$ (or their G^F-classes) are of the same **genus** if $C_G(s)$ and $C_G(t)$ are conjugate by an element of G^F. In $G^F = \mathrm{SL}(2,q)$, for example, the elements ± 1 of the center determine one genus; the remaining semisimple elements are regular and lead to two genera as in 8.8, corresponding to split and nonsplit maximal k-tori.

To get started, one can use the criteria of Carter (2.12) and Deriziotis (2.15) to work out the possible centralizer types in G. When $s \in G^F$, $C_G(s)$ is the (almost direct) product of a central torus and a semisimple group. Their fixed point subgroups have computable orders, whose product is the order of $C_G(s)^F$, thanks to the fact that isogenous groups over a finite field have the same number of rational points (B16.8). Carter develops the general theory in [**27**] and goes on to work out the details for classical groups in [**28**].

Deriziotis [**37, 38, 39, 40**] shows how to parametrize semisimple classes and their centralizers in a more visualizable way, by means of the "Brauer complex". This is a fundamental domain (alcove) for the affine Weyl group partitioned into q^r smaller copies of this alcove, each with a distinguished point in its closure representing one of the q^r semisimple classes of G^F. The organization of classes and centralizers by genus becomes transparent in this set-up. The work of Deriziotis includes a determination of centralizers for exceptional groups, which is extended to twisted groups by Deriziotis and Liebeck [**44**].

8.11. Counting Semisimple Classes in a Genus

The grouping of semisimple classes by genus is a natural step from the viewpoint of character theory. Long before the general theory described above was worked out, it was clear that "generic" character tables could be written down (when p is good) for certain families of groups such as $\mathrm{SL}(2, q)$: there the calculation goes back to Schur (see [**83**]). It is observed that character values on classes belonging to the same genus behave coherently, as Lusztig's work eventually explains in detail.

Some early papers on exceptional groups, already cited in Chapter 7, include a determination of semisimple classes and centralizer orders anticipating the later theory. See Chang [**30**] and Enomoto [**52**] for G_2, Shoji [**149**] and Shinoda [**147**] for F_4, Mizuno [**130**] for E_6. Later work by Fleischmann and Janiszczak [**56, 57**] recovers (with some corrections) the calculations of Mizuno for E_6 and treats the complicated cases E_7 and E_8 (with the help of a computer). Here the results of Deriziotis [**39**] are also rechecked.

Many of these calculations are directed toward the determination of the total number of classes in G^F (see 8.17 below). For this one has to count the number of semisimple classes in each genus. The procedure is outlined clearly in [**57**]. Following Carter, one can parametrize a genus by a pair $(J, [w])$, where J is a proper subset of the vertex set $\widetilde{\Delta}$ of the extended Dynkin diagram (2.15). Here $[w]$ ranges over a set of conjugacy class representatives of $W^J := N_W(W_J)/W_J$, where W_J is the subgroup of W generated by the reflections corresponding to J. To get the appropriate centralizer one twists by w the reductive subgroup of G generated by a maximal torus T and root groups $U_{\pm\alpha}$ for $\alpha \in J$. The fixed point group under F is the product of a torus part and a (possibly twisted) group of Lie type determined by J and w. The order of the torus part is found by evaluating at q the characteristic polynomial of w acting on the quotient of $X(T) \otimes \mathbb{R}$ by the subspace generated by J.

The cardinality of the collection of semisimple classes of G^F belonging to the genus $(J, [w])$ is then of the form $f(J, [w])/|C_{W^J}(w)|$, where $f(J, [w])$ counts the number of $t \in T$ for which $w \cdot F(t) = t$ and the subgroup of W fixing t is W_J. This is in fact a polynomial in q. To determine the $f(J, [w])$ explicitly is the hard part of the calculation, requiring Möbius inversion based on a collection of closed subsystems of Φ.

8.12. Regular Unipotent Elements

Next we consider the unipotent elements of G^F. We know from 4.6 that the regular unipotent elements in G form a single conjugacy class. Since a unipotent element is regular if and only if lies in a unique Borel subgroup, it is clear that F stabilizes the regular unipotent class. In turn, G^F must contain a regular unipotent element u (8.4). When p is good for G, u lies in $C_G(u)^\circ$ by 4.7. If moreover $Z(G) = e$, we know that $C_G(u) = C_G(u)^\circ$. Using these facts and Theorem 8.5, one can show:

THEOREM. *G^F contains a regular unipotent element of G. Moreover:*

(i) *The number of regular unipotent elements in G^F is $|G^F|/q^r$.*

(ii) *Assume $Z(G) = e$. If $u \in G^F$ is a regular unipotent, lying in an F-stable maximal unipotent subgroup U of G, the class of u in G intersects G^F in $|C_{U^F}(u)|/|C^\circ_{U^F}(u)|$ classes.*

(iii) *Assume that p is good for G, and that $Z(G) = e$ (or, more generally, that G is a connected reductive group with connected center). Then the regular unipotent elements lying in G^F form a single conjugacy class.*

The idea of (i) is easiest to sketch when G is split over k, but the argument is similar in the quasi-split case. If $B = TU$ is an F-stable Borel subgroup, the regular unipotents in B^F lie in U^F and are characterized in the factorization $\prod_{\alpha>0} x_\alpha(c_\alpha)$ by the condition that $c_\alpha \neq 0$ whenever α is simple (4.1). So B^F contains q^{m-r} regular unipotents, where $m := \dim U$ is the number of positive roots. Now each regular unipotent element of G lies in a unique Borel subgroup. If B is F-stable, the total number of F-stable Borel subgroups is $|\mathcal{B}^F| = |(G/B)^F| = |G^F|/|B^F|$ (as remarked in 8.4). In the split case, this number is $|G^F|/(q-1)^r q^m$. From this (i) follows. (Compare this with the discussion in 8.14 below of the total number of unipotent elements, coupled with the fact that $|G^F|$ is a polynomial in q of degree $r + 2m$.)

For more details, see Springer–Steinberg [**180**, III, 1.19–1.20], Digne–Michel [**46**, 14.23–14.25].

When p is bad the centralizer of a regular unipotent element fails to be connected, leading to more than one class in G^F.

8.13. Character Degrees

Assume that $Z(G)$ is trivial and p is good, so the regular unipotent elements lying in G^F form a single class (8.12). If u is such an element and k is an integer not divisible by p, u^k generates the same cyclic group as u and is therefore again regular. It follows that u and u^k are conjugate in G^F. This has an interesting consequence for the character theory of G^F, as shown by Green–Lehrer–Lusztig [**63**] (see also Carter [**29**, 8.3]).

Say $\rho : G^F \to \mathrm{GL}(V)$ is a representation over $\mathbb{C}$. The representing transformations are of finite order in $\mathrm{GL}(V)$ and hence semisimple. All eigenvalues involved are roots of unity, lying in some cyclotomic extension F of $\mathbb{Q}$. Under our assumptions, $\rho(u)$ and $\rho(u^k)$ are similar in $\mathrm{GL}(V)$ for all k not divisible by p. This amounts to saying that the Galois group of $F/\mathbb{Q}$ fixes the characteristic polynomial of $\rho(u)$, i.e., that the coefficients lie in $\mathbb{Q}$ (and hence in $\mathbb{Z}$, being algebraic integers). Because $\rho(u)$ is semisimple, some elementary reasoning shows that the (multiplicity-free) minimal polynomial also has coefficients in $\mathbb{Z}$, as do the other invariant factors which determine the rational canonical form. It follows that $\rho(u)$ is similar to an integral matrix. In other words, $\rho(u)$ stabilizes a $\mathbb{Z}$-form L of the vector space V. After reduction modulo p, we get a unipotent operator on L/pL, whose trace is equal to the dimension of V modulo p. Thus

$$\mathrm{Tr}\,\rho(u) \equiv \dim V \pmod{p}.$$

Having made this elementary observation, Green, Lehrer, and Lusztig go on to show that the values at u of irreducible characters of G^F are severely limited: only $0, 1$ or -1 can occur. Their argument combines a number of known facts about characters of G^F, involving induced characters from parabolic subgroups and from a Sylow p-subgroup. To summarize:

THEOREM. *Assume that $Z(G) = e$ (or that G is a connected reductive group with connected center) and that p is good for G. If u is a regular unipotent element in G^F, the value of any irreducible character of G^F at u is one of the integers $0, 1, -1$. Thus the degree of any irreducible character is congruent modulo p to $0, 1$, or -1.*

The last statement of the theorem is of course true when p is 2 or 3, even if this prime happens to be bad for G. But the assumption on the center of G is certainly needed, as the example of $\mathrm{SL}(2, q)$ (p odd) already shows [**83**]. Here one finds the trivial character, the Steinberg character (of degree q), families of irreducible characters of respective degrees $q+1$ and $q-1$ (all in harmony with the theorem); but one also finds two characters of degree $(q+1)/2$ and two of degree $(q-1)/2$. For $\mathrm{GL}(2, q)$ or $\mathrm{PGL}(2, q)$ these last four characters would no longer occur.

8.14. The Number of Unipotent Elements

Steinberg [**187**, §15] found a way to calculate the number of unipotent elements in any finite group G^F, by exploiting a particular irreducible complex character discovered by him and now known as the *Steinberg character.* (See Carter [**29**, Chap. 6] or the survey [**85**].) This character χ takes on easily computable integer values; only their absolute values are needed here. If $s \in G^F$ is semisimple, $Q(s) := |\chi(s)|$ is the power of p dividing the order of the centralizer of s in G^F. For all other $x \in G$, $\chi(x) = 0$.

THEOREM. *The number of unipotent elements in G^F is $q^{|\Phi|}$, the square of the order of a Sylow p-subgroup.*

It is easy to sketch the proof. Begin by counting the elements in G^F according to the Jordan decomposition: $|G^F| = \sum P(s)$, where the sum is over semisimple elements s and where $P(s)$ denotes the number of unipotent elements commuting with s. When $s \in Z(G^F)$, we are trying to compute $P(s)$. Next express $|G^F|$ in a different way. Since χ is an irreducible character, one of the standard orthogonality relations shows that its inner product with itself is 1. From the above description of character values, this translates into: $|G^F| = \sum Q(s)^2$, sum taken over all semisimple elements s. Now we have two expressions for $|G^F|$, which can be equated:

$$\sum P(s) = \sum Q(s)^2.$$

What has to be proved is that $P(s) = Q(s)^2$ for $s \in Z(G^F)$.

We can use induction on $\dim G$. When $s \in G$ is semisimple, we know (see 2.2) that even if the reductive group $C_G(s)$ fails to be connected, all its unipotent elements lie in the identity component and therefore in the (semisimple or trivial) derived group of the identity component. So if $s \notin Z(G^F)$, the induction hypothesis shows that $P(s) = Q(s)^2$, and all corresponding terms can be removed from the displayed sums. After dividing the remaining equation by $|Z(G^F)|$, we are done.

Note that the theorem is what one might naively guess, based on the fact that the unipotent variety of G is defined over k and has dimension $|\Phi|$ (see 4.2). But the unipotent variety is highly nonsingular, so there is no obvious way to read off the number of k-rational points.

Remark. Steinberg's approach to counting unipotent elements, though very elegant, is not the only possible one. Lehrer [**108**] has used combinatorial arguments, avoiding the Steinberg character, to show that both the number of unipotent elements of G^F (or F-stable nilpotent elements in $\mathfrak{g}$) and the number of F-stable maximal tori in G (or F-stable maximal toral subalgebras of $\mathfrak{g}$) are equal to $q^{|\Phi|}$. But there is no apparent direct link between the two calculations.

8.15. Unipotent Classes: Strategies

How can we compute the number of unipotent classes in G^F, and if possible exhibit explicit representatives for them? Two logically independent approaches might be entertained:

(1) Begin with the known facts about unipotent classes in G (as discussed in Chapter 7); these may require that p be "sufficiently large". Determine which of the finitely many unipotent classes in G are F-stable—or equivalently, by Theorem 8.4, meet G^F. For each such class in G, compute the component group $A(u) = C_G(u)/C_G(u)^\circ$ of an element u in the class, together with the way F acts on it. Thanks to 8.5, the number of F-conjugacy classes in $A(u)$ is the number of classes in G^F meeting the given class in G. With luck one may be able to find class representatives in G of the form $\prod_{\alpha>0} x_\alpha(c_\alpha)$ with c_α in the prime field, making the results more uniform. But of course one has to expect some differences when p is a bad prime.

(2) Start from scratch with the Bruhat decomposition in G^F and the Chevalley commutation relations in U^F (in the split case) or Steinberg relations (in the quasi-split case). Since U^F is a Sylow p-subgroup of G^F, it is enough to investigate the conjugacy classes of elements of the form $\prod_{\alpha>0} x_\alpha(c_\alpha)$. Select a list of candidates for class representatives (using any method, even guesswork). For each element on the list, compute directly the order of its centralizer in G^F: this may of course take a lot of work. Where the orders differ, the given elements cannot be conjugate; if the orders coincide, argue in other direct ways that conjugacy is impossible. Compute the total number of elements in the classes of these candidates: if it agrees with Theorem 8.14, the list is complete.

What commends method (2) is its directness—and the fact that it is reasonably effective for the exceptional types. But it lacks a natural starting point. On the other hand, method (1) provides a lot of detailed concrete information to get things started, and leads in practice to class representatives over the prime field, providing a smooth transition to finite fields. But it requires a large amount of preliminary work in the Lie algebra framework, involving the restriction $p \geq 3(h-1)$ (or Pommerening's substitute arguments).

In view of this, there is a tendency in the literature to use a hybrid approach, combining features of both methods (1) and (2). The Dynkin classification suggests what the class representatives might look like, after which the direct approach via the Bruhat decomposition and commutation relations in G^F can be used to compute centralizer orders and class sizes. Theorem 8.14 may provide a shortcut: once the correct number of unipotent elements has been accounted for, the list of classes must be complete. In the process the component groups $A(u)$ are sometimes determined independently by both methods. This kind of cross-check is reassuring, since there is no obvious *a priori* prediction of what the answers will be.

For classical types A–D, there are of course further possibilities—largely independent of the algebraic group approach—for determining unipotent classes by methods applicable to classical groups over arbitrary fields (of odd characteristic for types B–D).

8.16. Unipotent Classes: Exceptional Types

Here we summarize what is known about unipotent classes in types E, F, G, with indications of the methods used. Most of this work appeared between 1968 and 1980. As we have done earlier, we shall give more details for G_2.

(G_2) There is a single group G of type G_2: both simply connected and adjoint. Here the Coxeter number $h = 6$, so the theory surveyed in Chapter 7 can be applied uniformly when $p \geq 17$. From Theorem 8.5 together with the data on G_2 assembled in 7.12 and 7.18, we find that G^F has 7 unipotent classes. Indeed, all but one of the 5 classes in G intersects G^F in a single class, whereas the subregular class yields 3 classes: the Frobenius action on the component group S_3 is trivial, and there are 3 conjugacy classes in S_3. Unfortunately, these arguments break down for smaller p even though the result turns out to be the same whenever p is good (i.e., $p > 3$).

Chang [**30**] just assumes that $p > 3$ and works directly with the Bruhat decomposition and Chevalley commutation relations. He shows by explicit calculation that each element of the Sylow p-subgroup U^F is conjugate to one of the 7 exhibited elements in the table below and then computes their centralizers (which have different orders) to show that there are precisely 7 unipotent classes. Although Chang concentrates entirely on the finite groups, his calculations also confirm that G has 5 unipotent classes.

At the time Chang did his calculations, Steinberg's tally 8.14 of the total number of unipotent elements was not yet available; it would have cut down on the work. In any case, it is easy after the fact to check that Chang's list yields exactly q^{12} unipotents as required. To compute the class sizes here one just has to know that $|G^F| = q^6(q^2-1)(q^6-1)$.

G – class name	Dynkin weights	class representative	centralizer order
1	0 0	e	$\lvert G^F \rvert$
A_1	0 1	$x_\beta(1)$	$q^6(q^2-1)$
$\widetilde{A_1}$	1 0	$x_\alpha(1)$	$q^4(q^2-1)$
$G_2(a_1)$	0 2	$x_\beta(1)x_{2\alpha+\beta}(1)$	$6q^4$ or $2q^4$
		$x_\beta(1)x_{2\alpha+\beta}(\lambda)$	$2q^4$ or $6q^4$
		$x_\beta(1)x_{3\alpha+\beta}(\mu)$ or $x_\beta(1)x_{2\alpha+\beta}(-1)x_{3\alpha+\beta}(\zeta)$	$3q^4$
G_2	2 2	$x_\alpha(1)x_\beta(1)$	q^2

TABLE. Unipotent classes in $G_2(q)$, $p > 3$

In the table, some entries vary with the residue class of q modulo 3: the centralizer order for the fourth class is $6q^4$ when $q \equiv 1 \pmod 3$ and $2q^4$ when $q \equiv -1 \pmod 3$, with the reverse choice for the fifth class. The choice of representative for the sixth class also depends on whether $q \equiv 1$ or $q \equiv -1$ respectively. Several parameters have to be chosen in the field k of q elements: λ is a fixed nonsquare, μ is a fixed noncube, and ζ is a fixed element for which the polynomial $x^3 - 3x - \zeta$ is irreducible over k.

The missing primes $p = 2, 3$ are treated by Enomoto [**52**], using the same technique as Chang. In these cases the regular unipotent class of G splits into 2 (resp. 3) equal-sized classes in G^F: this illustrates the general fact for bad p that a regular unipotent element fails to lie in the identity component of its centralizer—indeed, its coset generates the component group (see 4.7). Otherwise there is little difference from Chang's list: the total number of unipotent classes is 8 when $p = 2$ and 9 when $p = 3$.

(F_4) Here there is again just one group G. Shinoda [**147**] determines the classes of G^F when $p = 2$, while Shoji [**149**] deals with the cases $p = 3$ and $p > 3$. Their method is essentially the same as that of Chang and Enomoto, but they take full advantage of Steinberg's theorem giving the total number of unipotents in G^F (here q^{48}). (Shoji also uses methods of Mizuno to study the other classes, thereby arriving at the total number of classes.)

As in the case of G_2, the explicit commutation relations require a fixed choice of signs for the structure constants in a Chevalley basis of the Lie algebra over $\mathbb{C}$ used to define the Chevalley group of type F_4: see Table 1 in Shoji [**149**]. The computation also requires detailed information about p-elements in subgroups of types B_3 and B_4. Along the way it is seen that component groups such as S_4 occur for centralizers of unipotents.

Ultimately the number of unipotent classes in G^F is found to be 35, 28, 26 when $p = 2$, $p = 3$, $p > 3$ respectively. In these respective cases the number of unipotent classes in G is seen to be 20, 16, 16 (in agreement with the Dynkin classification for sufficiently large p).

(E_r) Here the computations are truly formidable. In two lengthy papers [**130, 131**], Mizuno studies first E_6 and then E_7, E_8. In the cases of E_6 and E_7, there are distinct groups of simply connected and adjoint type. Moreover, there are quasi-split groups of type E_6 to be considered. These complications are offset somewhat by the fact that all roots have the same length, which considerably simplifies the commutation relations. As in previous work, the analysis begins with an explicit choice of signs for the structure constants in a Chevalley basis of each Lie algebra.

In [**130**] Mizuno treats the simply connected group G of type E_6. Dynkin's classification (with some "twisting" over the finite field) suggests a list of possible representatives for the unipotent classes of G as well as G^F, which he goes on to verify (with adjustments when $p = 2, 3$). As in previous work, this involves close study of the centralizers and leads to a determination of the component groups $C_G(u)/C_G(u)^\circ$. After considerable calculation, it turns out that (as expected from the classical theory) G has 21 unipotent classes when $p > 3$. The number of unipotent classes in G^F depends on the residue class of q modulo 3 (and differs for bad primes): for $p > 3$ there are 25 classes when $q \equiv -1$ and 39 when $q \equiv 1$; for $p = 3$ there are 27 classes; for $p = 2$ there are 28 classes when $q \equiv -1$ and 44 classes when $q \equiv 1$. Steinberg's Theorem 8.14 is again used to verify the completeness of the results: here there are q^{72} unipotent elements.

Similar (but even more complicated) calculations are carried out by Mizuno **[131]** in the cases of E_7 and E_8, leading again to explicit determinations of component groups of centralizers in G and to explicit class representatives in G^F.

Remark. Beynon and Spaltenstein **[11]** report some errors in Mizuno's summary tables, but not in the technical lemmas leading to these tables.

8.17. The Total Number of Classes

It is interesting to know the total number of classes in G^F: this coincides with the number of inequivalent irreducible representations of G^F over $\mathbb{C}$. We have already outlined the methods used to count semisimple and unipotent classes. One natural procedure for computing the total number of classes would be as follows. First assume that G is simply connected, so all centralizers of semisimple elements are connected.

(1) Assemble a list of representatives for the q^r semisimple classes of G^F, together with information about their centralizers.

(2) Choose for each such s a set of representatives u for the unipotent classes of $C_G(s)^F$. Since $C_G(s)$ is a connected reductive group of rank at most r, whose unipotent classes are those of its (semisimple) derived group, we can assume by induction on dimension that all required information on unipotent classes is at hand.

(3) Call $\mathcal{C}$ the set of all resulting ordered pairs (s, u). We claim that $\mathcal{C}$ is in natural bijection with the collection of all classes in G^F. Given $x \in G^F$, its semisimple part x_s is conjugate to a *unique* first coordinate s and its unipotent part x_u is then conjugate to at least one u for which $(s, u) \in \mathcal{C}$. This u is also unique: if x is conjugate in G^F to both su and sv, then u and v are clearly conjugate in $C_G(s)^F$. Now it follows that an element of G^F is conjugate to x precisely when it determines the same pair (s, u).

To make the counting process more efficient, one can group together all those semisimple classes in G^F for which the corresponding centralizers $C_G(s)$ are G^F-conjugate: each yields the same number of unipotent classes.

In case G fails to be simply connected, the groups $C_G(s)$ may fail to be connected. This obviously complicates the tally of unipotent classes. Finite group theorists are especially interested in the classes and characters of simple groups, so the groups G of adjoint type are the most natural starting point. To get around the connectedness problem, one can modify the steps above by counting the irreducible *characters* of G^F instead of the classes (these sets being of the same cardinality).

Lusztig's parametrization of characters allows one to get back to the semisimple classes of a simply connected group in duality (2.9) with the one being studied. As explained in [**29**, 12.9], there is a bijection between irreducible characters χ of G^F (up to equivalence) and pairs (χ_s, χ_u). Here χ_s is a "semisimple character" of G^F, parametrized by a semisimple class s^* of the dual group G^* fixed by a dual Frobenius map F^*. In turn, χ_u runs over the "unipotent" characters of the centralizer of s^* in $(G^*)^{F^*}$. Moreover, the degrees of these characters satisfy: $\chi(e) = \chi_s(e)\chi_u(e)$. The unipotent characters are quite subtle to study, and are not (apart from type A) precisely correlated with unipotent classes. But Lusztig has determined them—and their degrees—in all cases.

One or another version of the program just outlined has been carried out for many groups, especially the exceptional types (see the following section). A major ingredient is the determination of G^F-classes of centralizers of semisimple elements (8.11 above). Deriziotis [**41**] illustrates the method for type B_2 as well as the related Suzuki group in characteristic 2.

8.18. Class Numbers of Exceptional Groups

Here we summarize briefly the work done on class numbers of exceptional groups, which leads in each case to explicit polynomials in q with highest term q^r. The results vary somewhat within a given Lie type, depending on several additional factors:

- the isogeny type of G
- congruences satisfied by q
- whether the finite group is of Chevalley or twisted type
- whether p is good or bad

In addition to the work described in 8.16 above, Chang [**30**] and Enomoto [**52**] study semisimple classes and then arrive at the class number for groups of type G_2: when $p > 3$ there are $q^2 + 2q + 9$ classes; when $p = 2$ or 3, the number is $q^2 + 2q + 8$. (The explicit parametrization of classes here finds use in the later work of Chang–Ree and Enomoto on characters of G^F.)

Similarly, the work of Shinoda [**147**] and Shoji [**149**] (the latter corrected slightly by later calculations of Fleischmann and Janiszczak) leads to the following total number of classes in $F_4(q)$:

$$q^4 + 2q^3 + 7q^2 + 15q + 31 \text{ when } p > 3,$$

$$q^4 + 2q^3 + 7q^2 + 15q + 30 \text{ when } p = 3,$$

$$q^4 + 2q^3 + 6q^2 + 10q + 19 \text{ when } p = 2.$$

The groups of type E pose a much greater computational challenge, as is clear from Mizuno's extensive work on unipotent classes.

Deriziotis–Holt [**43**] work out the details for A_3 and ${}^2A_3$ as an example, and give computer-generated tables for adjoint groups of types E_6 and ${}^2E_6$ which lead to explicit class numbers (for all p including the bad primes 2, 3). The results differ slightly depending on the residue class of q modulo 6, so we give only a sample. When $q \equiv 1 \pmod 6$, the class number of adjoint $E_6(q)$ is

$$q^6 + q^5 + 2q^4 + 2q^3 + 9q^2 + 9q + 22,$$

while the class number of adjoint ${}^2E_6(q)$ is

$$q^6 + q^5 + 2q^4 + 4q^3 + 11q^2 + 11q + 16.$$

Fleischmann–Janiszczak [**56**] also use a computer to generate extensive tables for the adjoint groups of types E_6, ${}^2E_6$, and E_7. This leads to explicit determination of the class numbers, agreeing in the E_6 cases with the results of Deriziotis–Holt. When $p > 3$ the class number of $E_7(q)$ is found to be:

$$q^7 + q^6 + 2q^5 + 5q^4 + 13q^3 + 24q^2 + 47q + 59.$$

In [**57**] Fleischmann and Janiszczak go on to treat in a similar way the case of type $E_8(q)$. Here the generic class number (when $p > 5$) is found to be:

$$q^8 + q^7 + 2q^6 + 3q^5 + 10q^4 + 16q^3 + 40q^2 + 67q + 112,$$

with modifications for each of the bad primes $2, 3, 5$.

CHAPTER 9

Springer's Weyl Group Representations

In Chapter 6 we studied some aspects of the variety $\mathcal{B}_u$ of Borel subgroups containing a given unipotent element u. Springer [**174, 175**] found a way to make the Weyl group W act on the cohomology (ℓ-adic or classical) of $\mathcal{B}_u$, thereby introducing what are now known as **Springer representations** of W. These interact crucially with work of Deligne and Lusztig [**35**] on the representation theory of related finite groups G^F of Lie type. They also turn out to have interesting connections with other areas of Lie theory, notably the study of primitive ideals in the universal enveloping algebra of a semisimple Lie algebra over $\mathbb{C}$.

Our goal here is to provide a brief overview of Springer's theory and its application to G^F, with reasonably complete references to the large (and still growing) literature. As before, G denotes a connected semisimple group, assumed when convenient to be simple and of adjoint type, over the algebraically closed field K. Often K is taken to be $\mathbb{C}$ in the literature cited below; otherwise it should be of *good* prime characteristic, allowing one to pass freely from the unipotent variety $\mathcal{U}$ to the nilpotent variety $\mathcal{N}$ (6.20).

For other short accounts of the Springer representations, slanted respectively toward finite groups of Lie type and semisimple Lie algebras, see Carter [**29**, Chap. 12] and Collingwood–McGovern [**31**, Chap. 10]. A more extensive treatment is given by Shoji [**154**], emphasizing Lusztig's further work.

9.1. Representations of Weyl Groups

We are concerned here with representations of W over a splitting field of characteristic 0, which have been studied case-by-case in the simple types. In type A_r, W is the symmetric group S_{r+1}, whose irreducible representations and characters have been worked out in considerable detail by combinatorial and other methods (Frobenius, Schur, Young, Specht). In other classical types, W is an extension of a symmetric group by a 2-group (with the same W in types B and C), so similar methods may be used. The exceptional types have been studied mainly from the point of view of characters. (See Carter [**29**, 11.4, 13.2], Humphreys [**86**, 8.10].)

This case-by-case study reveals that the values of all irreducible characters of

Weyl groups lie in $\mathbb{Q}$. Even more is true: the irreducible representations themselves may be realized over $\mathbb{Q}$ (the Schur indices are all equal to 1). Springer's construction in [**175**] may be done over $\mathbb{Q}$ and thus provides a uniform explanation of this phenomenon.

Every Weyl group has at least two characters of degree 1: the trivial character 1 and the **sign character** ε taking values ± 1. (When W is a symmetric group, ε just gives the sign of a permutation.) If W is realized as an irreducible group generated by reflections, acting on an r-dimensional euclidean space E, the resulting irreducible character χ of degree r is called the **reflection character**.

One can obtain the regular representation of W by forming the quotient $\overline{S} := S/I$, where S is the symmetric algebra on E and I is the ideal generated by W-invariants of positive degree (see Bourbaki [**23**, V, 5.2]). The graded algebra $\overline{S}$ is called the **coinvariant algebra**. There is a canonical W-stable complement to I in S consisting of "harmonic" polynomials, which realizes the regular representation. Case-by-case study shows how the irreducible representations of W appear in the graded pieces here: see Steinberg [**182**], Lusztig [**115**, §2], Beynon and Lusztig [**9**]. For example, the trivial representation occurs in degree 0 and the sign representation occurs in the highest degree: the number of positive roots. This graded model of the regular representation (with degrees doubled) turns out to be a special case of Springer's construction, when the unipotent element is taken to be e. Here $\mathcal{B}_e = \mathcal{B}$, whose cohomology ring may be identified with $\overline{S}$ by a classical theorem of Borel.

The ideas in this chapter may be illustrated by a group G of type G_2, whose Weyl group is dihedral of order 12. If α and β are respective short and long simple roots, denote the corresponding reflections by $s := s_\alpha$ and $t := s_\beta$. Then the classes of W are represented by $e, s, t, st, (st)^2, (st)^3 = -e$. There are six irreducible characters of W, four of degree 1 and two of degree 2. The resulting character table is given below. The reflection representation (with character χ)

	e	s	t	st	$(st)^2$	$(st)^3$
1	1	1	1	1	1	1
ε	1	-1	-1	1	1	1
τ	1	1	-1	-1	1	-1
$\varepsilon\tau$	1	-1	1	-1	1	-1
χ	2	0	0	1	-1	-2
$\tau\chi$	2	0	0	-1	-1	2

TABLE. Character table of W for type G_2

can be realized over $\mathbb{Z}$ by taking α and β as basis vectors in a euclidean space $\mathbb{R} \otimes X(T)$ where W acts as a reflection group. In this way s and t are represented by the respective matrices

$$\begin{pmatrix} -1 & 3 \\ 0 & 1 \end{pmatrix} \text{ and } \begin{pmatrix} 1 & 0 \\ 1 & -1 \end{pmatrix}.$$

Study of the coinvariant algebra $\overline{S}$ shows that the characters of the graded pieces in degrees 0 to 6 are (respectively) 1, χ, $\tau\chi$, $\tau+\varepsilon\tau$, $\tau\chi$, χ, ε.

9.2. The Springer Correspondence

For any finite group H, denote by $\widehat{H}$ or $H\hat{}$ the set of its irreducible characters over a splitting field of characteristic 0 such as $\mathbb{C}$. When convenient, this may also be viewed as the set of isomorphism classes of irreducible H-modules over such a field.

At the level of characters, Springer obtains a bijection (called the **Springer correspondence**) between $\widehat{W}$ and a subset $\mathcal{S}$ of the G-classes of pairs (u, ϕ), where $u \in G$ is unipotent and ϕ is an irreducible character of the finite group $A(u) := C_G(u)/C_G(u)^\circ$. Two pairs (u, ϕ) and (v, ψ) are in the same G-class just when u is conjugate to v and ϕ corresponds to ψ under the resulting isomorphism of $A(u)$ with $A(v)$. Topologically, the pair (u, ϕ) corresponds to an irreducible, G-equivariant local system on the class of u, with the monodromy given by ϕ.

The character of W determined by the pair (u, ϕ) is denoted by $\chi_{u,\phi}$. Every unipotent class yields at least one character of W, but in general not all characters of $A(u)$ occur. Write $A(u)\hat{_0}$ for the set of those which do.

The reader should be cautioned that notation for the objects involved in the Springer correspondence varies throughout the literature, while the unipotent element u is often replaced by a nilpotent element of $\mathfrak{g}$. Moreover, the current convention (which we follow) is to multiply Springer's original characters of W by the sign character ε. With this convention, there are two extreme cases: $\chi_{e,1} = \varepsilon$ when $u = e$, whereas $\chi_{u,1} = 1$ when u is regular unipotent. (Note that $C_G(u)$ is connected and thus $A(u)$ is trivial in both of these cases when G is of adjoint type, char K being good.)

The actual irreducible representations of W affording the Springer characters are rather elusive. We may ask in general: how can one explicitly construct representations of a finite group H? If H acts as a group of permutations of a set, or as a group of automorphisms of a geometric object X (say an algebraic variety or topological space), the induced linear action of H on a space of functions on X may yield an interesting representation. Alternatively, the action of H on X may yield a linear action on associated cohomology groups $\mathrm{H}^i(X)$. Representations constructed in these ways, including induced representations, unfortunately tend to be far from irreducible.

Springer's cohomological construction of Weyl group representations is appreciably more subtle, since it does not arise so directly from an action of W on a variety. But in compensation one gets irreducible representations. Take $K = \mathbb{C}$ or K of good prime characteristic. Springer finds a linear action of W on the cohomology (classical or ℓ-adic) of the projective variety $\mathcal{B}_u$. Later constructions have involved other types of homology or cohomology (see 9.5 below). For the moment, we just write $\mathrm{H}^i(\mathcal{B}_u)$ for the resulting W-modules.

All of these constructions share certain basic features. Set $d(u) := \dim \mathcal{B}_u = (\dim C_G(u) - r)/2$ (see 6.17). One finds that the highest nonvanishing degree of cohomology is $2d(u)$. This is the real dimension of $\mathcal{B}_u$ if we are working over $\mathbb{C}$. Moreover, the dimension of $\mathrm{H}^{2d(u)}(\mathcal{B}_u)$ is the number of irreducible components of $\mathcal{B}_u$: there is a canonical basis corresponding to the set of components. In this degree one finds an irreducible constituent $\rho_{u,\phi}$ with character $\chi_{u,\phi}$.

What role does ϕ play here? The group $C_G(u)$ acts on $\mathcal{B}_u$ by conjugation, permuting the irreducible components (all of which have the same dimension). Each component is fixed by $C_G(u)^\circ$, thereby inducing a permutation action by $A(u)$. This in turn yields a natural linear action of $A(u)$ on each $\mathrm{H}^i(\mathcal{B}_u)$, commuting with the W-action defined by Springer. Consider the isotypic component of ϕ in $\mathrm{H}^{2d(u)}(\mathcal{B}_u)$: the sum of all $A(u)$-submodules with character ϕ. It is stable under the W-action and turns out to be a direct sum of copies of a single irreducible W-module. This defines a representation $\rho_{u,\phi}$ with the desired character $\chi_{u,\phi}$. More precisely, the work of Springer shows:

THEOREM. *With notation as above, let $u \in G$ be unipotent. Then:*

(i) *For each $\phi \in A(u)\hat{}$, the isotypic component of $\mathrm{H}^{2d(u)}(\mathcal{B}_u)$ corresponding to ϕ is zero or is an irreducible $A(u) \times W$-module realizing a representation $\phi \otimes \rho_{u,\phi}$. Here $\rho_{u,\phi}$ is an absolutely irreducible representation of W. Call its character $\chi_{u,\phi}$.*
(ii) *The assignment $(u, \phi) \mapsto \chi_{u,\phi}$ induces a bijection between a subset $\mathcal{S}$ of the G-classes of pairs (u, ϕ) and $\widehat{W}$. Here u ranges over all unipotent elements of G and ϕ ranges over a set $A(u)\hat{_0}$ of irreducible characters of $A(u)$ which always contains the trivial character $\phi = 1$.*
(iii) *Taking cohomology with rational coefficients, the representations $\rho_{u,\phi}$ may be realized over $\mathbb{Q}$.*

9.3. Example: G_2

Before dealing further with the methods used to construct and study Springer's representations, we look at the details of the Springer correspondence in the case of a group of type G_2 when char K is good (see Springer [**174**, 7.16]). Here there are five unipotent classes: see 7.12 and 7.18. The group $A(u)$ is trivial unless u is subregular, in which case $A(u) \cong S_3$. Taking into account the three irreducible characters of S_3, we thus have seven potential G-classes of pairs (u, ϕ). But the dihedral group W of order 12 has just six irreducible characters. It turns out that the pair involving a subregular element and the sign character of S_3 fails to occur in $\mathcal{S}$.

As in 9.1, label the characters of W as $1, \varepsilon, \tau, \varepsilon\tau, \chi, \tau\chi$. The table below lists the characters $\chi_{u,\phi}$ occurring in $\mathrm{H}^{2d(u)}(\mathcal{B}_u)$.

class name	class dim	$\dim \mathcal{B}_u$	$A(u)$	$\chi_{u,\phi}$
1	0	6	1	ε
A_1	6	3	1	$\varepsilon\tau$
$\widetilde{A_1}$	8	2	1	$\tau\chi$
$G_2(a_1)$	10	1	S_3	χ $(\phi = 1)$
				τ $(\deg \phi = 2)$
G_2	12	0	1	1

TABLE. Springer correspondence for type G_2

To see what is going on in the subregular case, recall from 6.11 and 6.12 that $\mathcal{B}_u$ is a Dynkin curve, consisting of a single line of type β intersected by three lines of type α. The line of type β is fixed by $C_G(u)$, while the lines of type α are permuted transitively. From this it follows that the permutation action of $A(u) \cong S_3$ on the four irreducible components of $\mathcal{B}_u$ has one fixed point together with three points permuted according to the natural action of S_3. But the 3-dimensional permutation representation of S_3 has a fixed line and an irreducible 2-dimensional complement. Putting this together, we find that $H^2(\mathcal{B}_u)$ splits into two isotypic $A(u)$-submodules of dimension 2. Springer's argument shows that each of these $A(u)$-submodules is irreducible for $A(u) \times W$. They turn out to realize the representations $1 \otimes \chi$ and $\phi \otimes \tau$, where $\deg \phi = 2$.

9.4. Other Simple Types

As in the case of G_2, the Springer correspondence has been worked out completely for all simple algebraic groups. To write down the details, one has to label the irreducible characters of W in some systematic way: see Carter [**29**, 13.2]. Similarly, one has to supplement the Bala–Carter notation for unipotent classes with labels for the various irreducible characters of $A(u)$ when this group is nontrivial. For an adjoint group of type A, $A(u)$ is always trivial. For other classical types it may be an elementary abelian 2-group. For exceptional types, it can only be a symmetric group S_k $(1 \leq k \leq 5)$. Thus the characters of $A(u)$ pose no serious problem. Moreover, all of them are afforded by representations over $\mathbb{Q}$, leading to similar behavior for the characters of W.

Carter [**29**, 13.3] writes down the Springer correspondence explicitly for each simple type. In type A_r, partitions of $r+1$ parametrize both the characters of $W \cong S_{r+1}$ and the unipotent classes; this gives the correspondence. (As pointed out at the end of 6.15, the number of irreducible components of $\mathcal{B}_u$ is equal to the degree of the irreducible character of the symmetric group labelled by the partition of u.) The other classical types are worked out by Shoji [**150**]. Even though the Weyl groups of types B and C are isomorphic, the numbers of unipotent classes in odd orthogonal and symplectic groups fail to match. So the Springer correspondence differs for these groups. Shoji [**151**] also deals with groups of type F_4. Types E_6, E_7, and E_8 are treated by Alvis, Lusztig, and

Spaltenstein in [**3**].

Some of the methods used for these calculations will be discussed in 9.9 below, in the wider context of determining all Springer representations in the $\mathrm{H}^i(\mathcal{B}_u)$.

9.5. Constructions of Springer Representations

Looking at the myriad approaches to Springer's representations in the literature cited below, one is at first reminded of the proverbial blind men attempting to describe an elephant. But eventually a coherent picture emerges. For example, the characters ϕ of $A(u)$ involved in defining the representations are seen to correspond geometrically to certain local systems on unipotent classes or nilpotent orbits.

Springer's constructions (see [**174, 175, 179**]) are carried out in two settings. If char $K = p > 0$, and ℓ is a prime different from p, one can study the ℓ-adic cohomology groups of the projective varieties $\mathcal{B}_u$. These are vector spaces over an algebraic closure of $\mathbb{Q}_\ell$. If on the other hand $K = \mathbb{C}$, then one can instead view $\mathcal{B}_u$ as a complex manifold and use classical cohomology theory. As a consequence of Chow's Theorem, $\mathcal{B}_u$ is still connected in the complex topology, with the same irreducible components as before: see for example Mumford [**132**, (4.16)] and [**133**, I, §10].

Kazhdan and Lusztig [**95**] offer a more elementary topological construction of Springer's representations $\rho_{u,\phi}$, by realizing them in the top Borel–Moore homology group $\mathrm{H}_{2d(u)}(\mathcal{B}_u)$.

Slodowy [**155**] proposes another construction of Springer representations, from the viewpoint of singularity theory. In studying Springer's desingularization of $\mathcal{U}$ in the context of a "simultaneous resolution" with fibres $\mathcal{B}_x$ for all $x \in G$, he is led to construct a monodromy representation of a certain fundamental group having W as a quotient. This ultimately comes from the action of W on the regular semisimple set in G, but yields an action of W on $\mathrm{H}^*(\mathcal{B}_u)$.

The heavier machinery of intersection cohomology, as developed by Deligne–Goresky–MacPherson, is brought into play by Borho–MacPherson [**21, 22**] and Lusztig [**117, 119**]. This yields decisive information about the representations of W occurring in the various $\mathrm{H}^i(\mathcal{B}_u)$, to be discussed below in 9.6. Further developments in this direction are due to Borho–Brylinski–MacPherson [**18, 19**] (emphasizing equivariant K-theory and characteristic classes), Ginsburg [**61**], and Kashiwara–Tanisaki [**92**].

Approaching the subject from a very different direction, Joseph [**91**] constructs representations of W in the setting of a complex semisimple Lie algebra $\mathfrak{g}$. Say $\mathfrak{b} = \mathfrak{h} \oplus \mathfrak{n}$ is a Borel subalgebra, with $\mathfrak{h}$ a Cartan subalgebra and $\mathfrak{n}$ the nilradical. If O is any nilpotent orbit in $\mathfrak{g}$, consider the irreducible components of $O \cap \mathfrak{n}$. To each of these Joseph attaches a polynomial function on the dual space $\mathfrak{h}^*$. Their span is then a W-submodule in the space of polynomials. For example, the zero orbit yields the trivial representation of W. The motivation behind all of this comes from the study of primitive ideals in the universal enveloping algebra $U(\mathfrak{g})$. (See Borho–Brylinski–MacPherson [**19**, §5], Collingwood–McGovern [**31**, Chap. 10], Jantzen [**88**, Kap. 14].)

Hotta [**75, 76, 77**] shows that Springer's representations (multiplied by the sign representation) may be identified with those constructed by the various methods just sketched. In doing so, he develops a "local" formula for the action of a simple reflection on the fundamental cycles in Springer's top degree representation. Vergne [**197**] makes the connection with Joseph's work more directly.

9.6. Properties of the Representations

The various approaches to the construction of Springer representations on cohomology spaces $\mathrm{H}^i(\mathcal{B}_u)$ lead to a reasonably complete description. Here we summarize briefly some of the main conclusions. While we have tried to attribute results correctly, there are by now quite a few alternative proofs in the literature.

(1) In most of the cohomology constructions it is seen very soon that

$$\mathrm{H}^i(\mathcal{B}_u) = 0 \text{ whenever } i \text{ is odd.}$$

This typically results from the existence of a paving of $\mathcal{B}_u$ by affine spaces, the prototype being the Bruhat decomposition of $\mathcal{B}$. (A variety X is said to be **paved by affine spaces** if X has a sequence of closed subsets $X_0 \subset X_1 \subset \cdots \subset X_d = X$, with each nonempty $X_k - X_{k-1}$ being a disjoint union of finitely many affine spaces of dimension k). See for example the treatments of $\mathrm{GL}(n, K)$ and other groups over fields of prime characteristic by Spaltenstein [**157**], Hotta–Springer [**80**, §2], Springer [**178**, §5].

(2) De Concini, Lusztig, and Procesi [**33**] look more closely at the characteristic 0 setting. If $n \in \mathfrak{g}$ is nilpotent, they show that *the integral homology of $\mathcal{B}_n$ is zero in odd degrees, and is torsion-free in even degrees.* Moreover, *all the homology comes from algebraic cycles.*

(3) When $u = e$, we have $\mathcal{B}_u = \mathcal{B}$. Its classical cohomology has been described explicitly by Borel (see [**12**] or [**13**, §20], as well as Borho–Brylinski–MacPherson [**19**, 1.5–1.6, 4.1–4.2]), and may be identified with the coinvariant algebra of W described at the end of 9.1 above. Here W acts according to the regular representation, suitably graded.

(4) Working in the setting of ℓ-adic cohomology, Hotta–Springer [**80**] show that the embedding of $\mathcal{B}_u$ in $\mathcal{B}$ induces a W-equivariant mapping $\mathrm{H}^*(\mathcal{B}) \to \mathrm{H}^*(\mathcal{B}_u)$. This map respects the gradings, and has image equal to the fixed point space $\mathrm{H}^*(\mathcal{B}_u)_1$ of $A(u)$: the ϕ-isotypic subspace when $\phi = 1$. Moreover, $\mathrm{H}^{2d(u)}(\mathcal{B}) \to \mathrm{H}^{2d(u)}(\mathcal{B}_u)_1$ is surjective.

(5) The intersection cohomology construction of Borho–MacPherson [**21**] gives the most precise information about the multiplicities of representations $\rho_{u,\phi}$ in

$\mathrm{H}^i(\mathcal{B}_v)$ for unipotent elements u, v:

$$\text{If } \rho_{u,\phi} \text{ occurs in } \mathrm{H}^i(\mathcal{B}_v), \text{ then } \mathrm{Cl}(v) \subset \overline{\mathrm{Cl}(u)} \text{ and } i \geq 2d(u).$$

The exact multiplicities are given here in terms of local intersection cohomology. Moreover:

$$\text{If } i < 2d(u), \text{ then } \rho_{u,\phi} \text{ does not occur in } \mathrm{H}^i(\mathcal{B}_u).$$

(6) The results just summarized enable one to pin down the representations $\rho_u := \rho_{u,1}$, which play a crucial role in Lusztig's work on characters of G^F as well as work of Barbasch and Vogan on Lie group representations. Alvis and Lusztig [**3**] determine in a precise way the subset of $\widehat{W}$ consisting of all ρ_u. This involves induction of unipotent classes in the sense of Lusztig–Spaltenstein [**123**] (see 5.10), along with induction of representations from subgroups of W corresponding to proper subsets of the vertices of the extended Dynkin diagram. Among the representations ρ_u are those which Lusztig calls "special". These are in bijection with the irreducible "unipotent" representations of G^F: see Carter [**29**, 11.3, 12.6]. (There is in turn a bijection with "special" unipotent classes of G, where a duality on the closure ordering is described by Spaltenstein [**160**, Chap. III].)

(7) Direct study of $\mathrm{GL}(n, K)$ provides alternative proofs of many of the general results about Springer representations (and Green polynomials), or yields finer detail. For example, over $\mathbb{C}$ the map $\mathrm{H}^*(\mathcal{B}) \to \mathrm{H}^*(\mathcal{B}_u)$ is surjective. Going over to homology, this means that $\mathrm{H}_*(\mathcal{B}_u)$ embeds in $\mathrm{H}_*(\mathcal{B})$. A precise description of the embedding is given by Varagnolo [**196**] in terms of harmonic polynomials, thus recovering Springer's representations of symmetric groups. Among the papers treating $\mathrm{GL}(n, K)$ are Springer [**174**], Spaltenstein [**157**], Hotta–Shimomura [**79**], Shimomura [**146**], De Concini–Procesi [**34**].

9.7. Generalized Springer Correspondence

A puzzling aspect of the Springer correspondence is the failure of some characters ϕ of $A(u)$ to appear in the pairs $(u, \phi) \in \mathcal{S}$. We saw this in the case of G_2, where the nontrivial character of degree 1 of S_3 plays no role when u is subregular.

Lusztig [**119**] has found a way to complete the picture. The missing pairs provide essential building blocks for his theory of "character sheaves", part of his program for computing the character values of finite groups of Lie type. He develops a generalized Springer correspondence, with a natural extension of the Borho–MacPherson results, leading to generalized Green functions. At the same time he is able to treat small prime characteristics, where the details of course differ when the numbers of unipotent classes or nilpotent orbits differ. See Lusztig–Spaltenstein [**124**] and Spaltenstein [**165**], as well as the detailed exposition by Shoji [**154**].

9.8. Green Functions

As indicated earlier, the character theory of finite groups of Lie type has been a primary motivation for the study of Springer representations. The story begins with the seminal work of Green [**62**] on the characters of $\mathrm{GL}(n,q)$. Here the partitions of n parametrize both the unipotent classes and the classes (or irreducible characters) of the Weyl group S_n. A key ingredient in Green's work is a family of polynomials in q indexed by pairs of partitions. These are now known as **Green polynomials**, and have been studied thoroughly from the combinatorial viewpoint: see Macdonald [**125**, III.7]. These polynomials give the values of certain characters on the unipotent set of $\mathrm{GL}(n,q)$, contributing to a recursive determination of all character values.

Green's ideas are generalized in a powerful way by Deligne–Lusztig [**35**], who construct (virtual) characters of finite groups of Lie type using ℓ-adic cohomology of certain associated varieties. For extensive treatments see Carter [**29**] and Digne–Michel [**46**]. Here the basic setting is that of a connected reductive group G over a finite field $\mathbb{F}_q$ of characteristic p; this permits an inductive treatment of Levi subgroups (beginning with maximal tori). Let F be the associated Frobenius map. To an F-stable maximal torus T and a complex character θ of T^F one associates a virtual character $R_{T,\theta}$. This is the alternating sum of characters obtained by making G^F act on certain ℓ-adic cohomology groups for a prime ℓ different from p.

It turns out that for unipotent elements $u \in G^F$, the value of $R_{T,\theta}(u)$ is independent of θ and lies in $\mathbb{Z}$. Denote it by $Q_{T,G}(u)$ or just $Q_T(u)$. The functions Q_T are the Deligne–Lusztig version of Green polynomials and are called **Green functions** (expected to be polynomials in q). The Green functions satisfy "orthogonality relations" and play a crucial role in Lusztig's overall program to calculate the character values of G^F.

Recall from 8.8 that the F-stable maximal tori of G fall into G^F-classes parametrized by F-conjugacy classes in W. Taking q to be large enough so that G is split over $\mathbb{F}_q$, we are just dealing with the classes of W. If we fix an F-stable T, representatives T_w of other classes are obtained by "twisting" by elements of W. Thus the values of Green functions may also be denoted by $Q_w(u)$. If we fix u but vary w, we get in this way class functions on W which should be virtual characters.

Springer [**174**] goes on to propose a different construction (still in the framework of ℓ-adic cohomology). In his approach, Green functions are defined as follows. Fix an F-stable maximal torus T in G, and twist by $w \in W$ to obtain T_w. For each nilpotent element $n \in \mathfrak{g}^F$ one gets ℓ-adic cohomology groups $\mathrm{H}^i(\mathcal{B}_n)$ on which W acts. There is also an induced action of F (written F^*). The Green function associated to w is the function on the nilpotent set of $\mathfrak{g}^F$ defined by:

$$Q_{T_w,G}(n) = \sum_{i\geq 0}(-1)^i \operatorname{Tr}(F^* w, \mathrm{H}^i(\mathcal{B}_n)).$$

(In some formulations, w is replaced by w^{-1} on the right side.) When p is good, $Q_{T_w,G}$ may instead be regarded as a function on the unipotent set of G^F. These Green functions are constant on G^F-orbits or classes. Assuming both p and q are "sufficiently large", Kazhdan [**93**] verifies that Springer's Green functions agree with those defined by Deligne and Lusztig. (The bounds on p and q are improved in later case-by-case work.)

When G is of (split) classical type, Srinivasan [**181**] shows that the Green functions are in fact *polynomials* in q. Further case-by-case study has extended this to all types, at least in good characteristic (see 9.9 below). The appearance of powers of q is due to the fact that the Frobenius operator F^* has eigenvalues on $\mathrm{H}^i(\mathcal{B}_n)$ of absolute value $q^{i/2}$. In the setting of ℓ-adic cohomology, this is a "purity" result: see Springer [**178**] (whose proof uses [**164**]).

9.9. Computations of Green Functions

As indicated in 9.3 and 9.4 above, the Springer correspondence between characters of W and G-classes of pairs (u, ϕ) has been worked out explicitly for each simple type. This involves only the top degree of cohomology for $\mathcal{B}_u$, whereas the computation of Green functions requires detailed examination of all degrees. This too has been achieved in most cases (explicitly or algorithmically), largely through the efforts of Shoji and of Beynon and Spaltenstein. However, some small values of p and q excluded by their methods have to be looked at separately.

Groups of types A and G_2 are already treated by Springer [**174**]. From the results for $\mathrm{GL}(n, K)$ or $\mathrm{SL}(n, K)$ one then gets results for unitary groups: see Hotta–Springer [**80**]. Shoji [**153**] develops an inductive algorithm for computing Green polynomials of the remaining classical groups in odd characteristic, similar to his procedure for F_4 in [**152**]. He makes full use of the Borho–MacPherson results, and shows moreover that characters ϕ not in $A(u)_0^{\wedge}$ do not occur in any $\mathrm{H}^i(\mathcal{B}_u)$. (Working from the Deligne–Lusztig definition of Green functions, Lusztig [**113**] had already obtained closed formulas for the Q_w in the special case when w is a Coxeter element in the Weyl group of a classical group, without any restriction on p.)

Using methods like those of Shoji (together with a computer), Beynon and Spaltenstein [**11**] go on to determine the multiplicities of irreducible characters of W in each $\mathrm{H}^i(\mathcal{B}_u)$, when G is of type E_r $(r = 6, 7, 8)$. In particular, they verify the vanishing of all cohomology in odd degrees. Their paper gives detailed tables for type E_6, together with selected results for E_7 and E_8 which are most useful for applications. Complete tables are assembled in [**10**].

As a byproduct of their work, they provide an independent verification of the results of Mizuno [**131**] on the closure ordering (7.19). Even though they find some errors in Mizuno's tables, their results agree with his lemmas. To describe how the Springer representations determine the closure ordering, let $u, v \in G$

be unipotent. Then (based on Borho–MacPherson [**21**]) the following conditions are equivalent:

(i) $\mathrm{Cl}(v) \subset \overline{\mathrm{Cl}(u)}$.
(ii) The representation $\rho_{u,\phi}$ occurs in $\mathrm{H}^i(\mathcal{B}_v)$, for some $i \geq 0$ and some $\phi \in A(u)\hat{_0}$.
(iii) The representation $\rho_{u,1}$ occurs in $\mathrm{H}^{2d(u)}(\mathcal{B}_v)$.

The reader should compare this with the table for type G_2 in 9.10 below, where the closure ordering is linear.

Most of the cited calculations of Green functions require that p be good for G and that q be "not too small". But the study of characters of G^F involves all possibilities, extending to the groups of Suzuki and Ree defined only in characteristic 2 or 3. By now many of the remaining gaps have been filled: see for example Deriziotis–Michler [**45**], Lusztig [**120**, 1.4], and Malle [**126, 127**].

As this quick survey indicates, the multiplicities of the irreducible representations $\rho_{u,\phi}$ in the $\mathrm{H}^i(\mathcal{B}_v)$ are known very explicitly for the exceptional types. For classical types the methods are more algorithmic in nature, leaving some questions unsettled. For example, their study of the topology of hyperplane complements leads Lehrer and Shoji [**109**] to look for the precise multiplicity of the reflection representation of W in the various Springer representations. Their partial results are completed by Spaltenstein [**168**], who is also able to handle the case of characteristic 2. He relies heavily on the methods of Borho–MacPherson, emphasizing the relationship between fixed points on $\mathcal{B}$ and on the generalized flag varieties $\mathcal{P}$ consisting of conjugates of a single parabolic subgroup. Other questions left open by Lehrer and Shoji have been studied by J.M. Douglass.

9.10. G_2 Revisited

In the G_2 case, we can expand the table in 9.3 to a full description of Green polynomials as in the table below (under the assumption that $p > 3$ and $q > 5$). Here the subregular class in G yields three classes in G^F.

class of u	$\dim \mathcal{B}_u$	$Q_w(u)$
1	6	$\varepsilon(w)(q^2-1)(q^6-1)/(q^2-\chi(w)q+\varepsilon(w))$
A_1	3	$\varepsilon(w)\tau(w)q^3+\tau(w)\chi(w)q^2+\chi(w)q+1$
$\widetilde{A}_1$	2	$\tau(w)\chi(w)q^2+(\tau(w)+\chi(w))q+1$
$G_2(a_1)$	1	$(\chi(w)+2\tau(w))q+1$
		$(\chi(w)-\tau(w))q+1$
		$\chi(w)q+1$
G_2	0	1

TABLE. Green polynomials for type G_2

In accordance with the discussion of the coinvariant algebra at the end of 9.1, the characters occurring in $\mathrm{H}^{2i}(\mathcal{B})$ for $0 \leq i \leq 6$ are (respectively):

$$1,\ \chi,\ \tau\chi,\ \tau+\varepsilon\tau,\ \tau\chi,\ \chi,\ \varepsilon.$$

All characters of W except τ have the form $\rho_{u,1}$. Among these five only three are "special" in Lusztig's sense: $\varepsilon, \chi, 1$. These correspond respectively to the trivial, subregular, and regular unipotent classes.

9.11. Affine Springer Representations

An affine analogue of Springer's theory has been developed by Kazhdan and Lusztig [**96**]. They work with the points of G and $\mathfrak{g}$ over a field F of formal power series with coefficients in $\mathbb{C}$. The set of "Iwahori subalgebras" of the resulting affine Lie algebra $\mathfrak{g}(F)$ is an infinite-dimensional algebraic variety: an increasing union of ordinary projective varieties over $\mathbb{C}$. But the set of Iwahori subalgebras containing a "nil-elliptic" element is a finite-dimensional variety: these special elements of nilpotent type avoid the obvious problem associated with an element as irregular as 0. The cohomology of this variety affords representations of the affine Weyl group associated to W. (Further work in this direction has been done in the U. Chicago thesis of D.S. Sage, using a Borel–Moore homology construction in the spirit of [**95**].)

As a byproduct of their work, Kazhdan and Lusztig define a map from the nilpotent orbits in $\mathfrak{g}$ to the conjugacy classes of W. This agrees for $\mathfrak{sl}(n, \mathbb{C})$ with the obvious map based on partitions of n, but is more complicated in other cases—and fails to be surjective. See Spaltenstein [**167, 169**].

CHAPTER 10

Complements

We conclude with a brief survey of some work closely related to the main themes of this book. Unless stated otherwise, G continues to denote a connected semisimple group over an algebraically closed field K of arbitrary characteristic.

10.1. Unipotent Classes and Cells in Affine Weyl Groups

Lusztig has discovered a subtle bijection between unipotent classes in G and "two-sided cells" in the associated affine Weyl group (when $\operatorname{char} K$ is good). In a series of papers culminating in [**121**] and other papers cited there, he studies affine Weyl groups and their partitions into "cells" (left, right, two-sided). This is an important part of the combinatorial study of Coxeter groups initiated by Kazhdan and Lusztig in their famous 1979 paper on representations of Hecke algebras [**94**] (see also [**86**, Chap. 7]). Around that time Lusztig conjectured the following result, which he succeeded in proving only after considerable further work:

THEOREM. *Let* $\operatorname{char} K = 0$. *Let G have Weyl group W and associated affine Weyl group W_a. Then there is a bijection between the unipotent classes in G and the two-sided cells in W_a. In this bijection, cells which are finite correspond to those classes of unipotent elements u for which $C_G(u)^\circ$ is unipotent.*

Along the way it is shown that $\dim \mathcal{B}_u$ (see 6.7) coincides with a combinatorially defined function on the corresponding two-sided cell of W_a: Lusztig's "a-function" computed in terms of the Kazhdan–Lusztig basis of the Hecke algebra. While the cells in affine Weyl groups are defined in an elementary way, the only known proof of the theorem involves much deeper information from the representation theory of affine Hecke algebras. In particular, essential use is made of Lusztig's theory of "character sheaves".

The bijection in the theorem can be carried over formally to semisimple groups over fields of good prime characteristic (see 6.20). But the methods involved in the proof are inherently based in characteristic 0.

The definition of two-sided cells leads to a natural partial ordering. On the other hand, unipotent classes in G are partially ordered in terms of closures

(7.19). Lusztig conjectures that the bijection in the theorem should be order-reversing, which has been verified in many special cases by J. Shi. In all cases, the identity element of W_a lies in a two-sided cell by itself, corresponding to the regular unipotent class of G.

10.2. Embedding Unipotents in Rank 1 Subgroups

When $\operatorname{char} K = p > 0$ (and p is good for G), the work surveyed in Chapter 7 provides a coherent classification of unipotent classes or nilpotent orbits. Testerman [**195**] has further exploited many of the methods involved, in order to determine when it is possible to find a subgroup of G isomorphic to $\mathrm{SL}(2, K)$ or $\mathrm{PGL}(2, K)$ containing a given unipotent element u. An obvious necessary condition is that $u^p = e$. This turns out (after considerable case-by-case investigation) to be sufficient as well:

THEOREM. *Let* $\operatorname{char} K = p$ *be good for* G. *If* $u \in G$ *satisfies* $u^p = e$, *then* u *lies in some subgroup of* G *isomorphic to* $\mathrm{SL}(2, K)$ *or* $\mathrm{PGL}(2, K)$.

As she points out, it is not known whether one actually has to require that p be good.

The theorem is in a way the best possible group-theoretic analogue in characteristic p of the Jacobson–Morozov Theorem. It has applications to the determination (by Seitz and Testerman) of the maximal closed subgroups of simple algebraic groups: in particular, it leads to a construction of maximal subgroups of type A_1 in exceptional groups when p is good. Moreover, the proof introduces some techniques for exponentiating Lie algebra elements which are of independent interest.

One important ingredient in the proof is a precise determination of the orders of Richardson elements (5.3) in distinguished parabolic subgroups. Say P is distinguished (7.13), and corresponds to a subset θ of the simple roots Δ. Express the highest root as a $\mathbb{Z}^+$-linear combination of Δ, and let m be the sum of the coefficients of elements of $\Delta - \theta$. Then *the Richardson elements in* P *have order equal to the least power of* p *exceeding* m. Combined with the Bala–Carter–Pommerening parametrization of unipotent classes, this yields a recursive method to calculate the orders of all unipotent elements in G.

For example, in type A_r, the only distinguished parabolics are the Borel subgroups, for which θ is empty and $m = r$. In a Borel subgroup the Richardson elements are the regular unipotents. If u is one of these, $n := u - e$ is nilpotent; the first power of n equal to 0 is clearly n^{r+1}. From the binomial expansion, we then see that the p-power q satisfies $u^q = e$ only when $q \geq r + 1$.

Testerman's proof of the order formula requires case-by-case calculation, with some computer assistance. In the exceptional cases, a Chevalley basis must be specified in the corresponding Lie algebra over $\mathbb{C}$, leading to explicit recipes for nilpotent and unipotent elements.

10.3. Parabolic Subgroups with Abelian Unipotent Radical

Richardson classes (5.3) arise from the action of a parabolic subgroup P on its unipotent radical V. In some important special cases, Richardson, Röhrle, and Steinberg [**141**] have obtained more detailed information about the way a Levi factor of P acts by conjugation on V.

Let G be a connected reductive group over K, with $\operatorname{char} K$ assumed to be different from 2 when two root lengths occur in a simple factor of G. When $P = LV$ is a parabolic subgroup with Levi factor L, the study of L-orbits in V is closely related to the study of orbits in G/P or (P, P) double cosets. Here the focus is mainly on the important special case in which V is abelian. When $K = \mathbb{C}$, the associated homogeneous spaces G/P are known to be precisely the simply connected Hermitian symmetric spaces of compact type, whose orbits under L and a related real form of G have been extensively studied by J.A. Wolf and others.

Using standard tools such as the Bruhat decomposition, along with detailed information about the individual root systems, the authors of [**141**] obtain sharp results. For example, each (P, P) double coset meets the unipotent radical V^- of an opposite parabolic subgroup in a single L-orbit; this yields a bijection between (P, P) double cosets and L-orbits in either V^- or V. These are also in natural bijection with (W_L, W_L) double cosets of W, where W_L and W are the respective Weyl groups of L and G. Along the way, explicit orbit representatives and orbit dimensions are found.

10.4. Conjugacy Classes of n-tuples

Working mainly in characteristic 0, Richardson [**140**] generalizes many of the standard results about the conjugation action of G on itself to the diagonal action of G on $G^n := G \times \cdots \times G$, or the derived action on $\mathfrak{g}^n$. The assumption on the characteristic is invoked to insure that reductive groups are linearly reductive, thus allowing Kempf's version of the Hilbert–Mumford Theorem to be exploited.

He begins by reviewing some features of the conjugation action in the $n = 1$ case. The orbit of $x \in G$ is closed (resp. unstable) precisely when x is semisimple (resp. unipotent). Given the Jordan decomposition $x = su$, the stabilizer (here the centralizer) G_x is the intersection of the stabilizers G_s and G_u, while the closed orbit (class) $G{\cdot}s$ is the unique closed orbit in the closure of $G{\cdot}x$. Moreover, the fibres of the adjoint quotient all have codimension r in G.

To generalize all of this to the action on G^n, consider a typical n-tuple $x := (x_1, \dots, x_n) \in G^n$. Define $A(x)$ to be the closed subgroup of G generated by $\{x_1, \dots, x_n\}$. Call x *semisimple* if $A(x)$ is linearly reductive and *unipotent* if $A(x)$ is unipotent. This agrees with the usual definitions when $n = 1$, though it leads to surprises when $n > 1$. Richardson shows that the orbit $G \cdot x$ is closed (resp. unstable) precisely when x is semisimple (resp. unipotent).

There is a reasonable substitute here for the Jordan decomposition. Fixing a Levi subgroup L of $A(x)$, whose unipotent radical is V, one can write each $x_i = y_i z_i \in LV$. Then $y := (y_1, \dots, y_n)$ and $z := (z_1, \dots, z_n)$ provide a (non-unique) "Levi decomposition" $x = y \cdot z$. The stabilizers in G behave as one might hope: $G_x = G_y \cap G_z$, while $G \cdot y$ is the unique closed orbit in the closure of $G \cdot x$. There is also a good analogue of the adjoint quotient, whose fibre dimensions can be calculated.

10.5. Centers of Centralizers

Springer's Theorem 1.14 says that $C_G(x)$ always contains a closed commutative subgroup of dimension at least r. Having seen in Chapter 2 that regular elements are those with centralizers of dimension exactly r, we concluded that $C_G(x)^\circ$ is commutative whenever x is regular. It is natural to ask about the converse. Kurtzke [**105, 106**] shows that, at least when $\operatorname{char} K$ is *good*, $C_G(x)^\circ$ is noncommutative whenever x is irregular. More precisely:

THEOREM. *Let* char K *be good for the reductive group* G. *Then for all* $x \in G$, $\dim Z(C_G(x)^\circ) \leq r$. *If equality holds, then* x *must be regular* (*unless* G *is of type* G_2).

In the case of $\mathrm{GL}(n, K)$ or $\mathrm{SL}(n, K)$, this comes from a classical matrix computation: for $x \in \mathrm{GL}(n, K)$ the center of the centralizer of x consists of precisely the polynomials in x which lie in $\mathrm{GL}(n, K)$. Thus the dimension of the center equals the degree of the minimal polynomial of x, at most n. We saw in 1.13 that equality holds just when x is regular (i.e., its minimal and characteristic polynomials coincide). In $\mathrm{SL}(n, K)$, both the rank and the dimension of the center drop by 1.

The proof of the theorem is easy when x is semisimple, thanks to Theorem 2.2. For arbitrary x, one has $C_G(x) = C_H(x_u)$ with $H = C_G(x_s)$ reductive. So the essential case is that of a unipotent element. Now the assumption of good characteristic permits a shift to the study of nilpotent elements in $\mathfrak{g}$. Here one invokes the combination of classical and newer methods surveyed in Chapter 7. Using Dynkin's work, there is a further reduction to the case of "semiregular" nilpotent elements in a regular semisimple subalgebra of maximal rank. Then case-by-case arguments come into play.

For exceptional types E and F, some computer programs are used to solve linear equations which determine the dimension of a space of matrices commuting with given ones. Naturally, a more conceptual proof would be preferable. But there really is an exception for G of type G_2: the center of the (4-dimensional) centralizer of a subregular unipotent element has dimension 2.

10.6. Torsion Primes

When G is simply connected, centralizers of semisimple elements are connected (Theorem 2.11). Steinberg [**189**] explores some closely related connectedness questions which require avoidance of "torsion" primes. These are somewhat analogous to bad primes, as defined in 3.9.

Following Springer–Steinberg [**180**, I, 4.1], a prime p is called a **torsion prime** for the irreducible root system Φ if the finitely-generated abelian group $L(\Phi^\vee)/L(\Phi_1^\vee)$ (quotient of coroot lattices) has p-torsion for some closed subsystem Φ_1 of Φ. It turns out that the only torsion primes are $p = 2$ (for all types except A_r and C_r), $p = 3$ (for types F_4, E_6, E_7, E_8), and $p = 5$ (for type E_8). When G is semisimple, the torsion primes for G are those attached to the various irreducible components of its root system Φ.

THEOREM. *Let G be a connected, simply connected semisimple group.*

(i) *Let s be a semisimple element such that $s^n \in Z(G)$ for some n divisible by no torsion prime. Then the derived group of $C_G(s)$ is also simply connected.*

(ii) *If* char K *is not a torsion prime, then $C_G(s)$ is connected for all semisimple elements $s \in \mathfrak{g}$.*

These conclusions are stated without proof in [**180**, II, 5.6, 3.19], having first been obtained around 1963. The motivation for part (ii) is obvious. Part (i) plays a role in the proof of (ii), as well as in the study of centralizers of sets of commuting semisimple elements in G. Note that Theorem 2.11 already implies that any pair s, t of commuting semisimple elements of the simply connected group G lie in some maximal torus: just take a maximal torus of the (connected!) group $C_G(s)$ containing t. But for three or more commuting elements this kind of argument breaks down, if the derived group of $C_G(s)$ fails to be simply connected. Elements of finite order are of special concern here, leading to the delicate formulation of (i).

The focus on torsion primes is suggested by earlier work of Borel on compact Lie groups, much of which can be recovered from Steinberg's more general arguments.

10.7. Classical Groups over Arbitrary Fields

Our viewpoint has been that of algebraic groups, mainly over an algebraically closed field. But classes and centralizers in the classical groups may also be studied directly, over almost any field or division ring (with some complications in characteristic 2). This is exemplified by the work of Wall [**198, 199**].

The first of these papers incorporates and extends a large amount of earlier work, especially that of J. Williamson. Two problems are addressed: (1) Attach suitable invariants to the classes of a unitary, symplectic, or orthogonal group over a division ring D (equipped with an involutory anti-automorphism). (2) Determine the equivalence classes of nondegenerate sesquilinear forms on finite-dimensional vector spaces over D. The first step is to show that (1) may be reduced to (2). This is done by associating to each element of the classical group in question a nondegenerate form, so that two elements are conjugate if and only if the associated forms are equivalent. The method may be viewed as a generalization of the Cayley parametrization for orthogonal groups. The actual classification of forms involves some further technical steps which we will not attempt to describe here. Suffice it to say that the results are sufficiently concrete to permit explicit calculations for classical groups over finite fields.

The later paper [**199**] concentrates instead on close relatives of the general linear group (over any field F): special linear groups, as well as projective general and special linear groups. Here the problem is to see how the well-known classes in $\mathrm{GL}(n, F)$ compare with classes in the related groups. The results are very explicit for finite F, bringing out some symmetry which helps to explain duality phenomena observed earlier by G.I. Lehrer and I.G. Macdonald.

10.8. Local and Global Fields

Consider an algebraic number field F and its completions F_v relative to discrete valuations v. If G is an algebraic group defined over F, with x conjugate to y in $G(F)$, then equally well x is conjugate to y in $G(F_v)$. We say that the classes in $G(F)$ satisfy the **Local–Global Principle** if conjugacy of $x, y \in G(F)$ in all $G(F_v)$ implies conjugacy in $G(F)$. Using the results of Wall [**198**], Asai [**4**] studies various classical groups (orthogonal, symplectic, unitary) over F, and proves that the Local–Global Principle is satisfied by their classes over F.

In turn, Bartels [**8**] looks at this question for arbitrary algebraic groups defined over F, in the framework of Galois cohomology. When G is defined over F, the **Hasse Principle** holds for G if for any finite Galois extension E/F, the natural mapping

$$\mathrm{H}^1(E/F, G(E)) \to \prod \mathrm{H}^1(E_w/F_v, G(E_w))$$

is injective, the product being taken over pairs v, w with w extending v. (This is known to be true, for example, when G is a simply connected semisimple group—though the stubborn case of type E_8 has been settled only in 1989 by V. Chernousov. See the survey by Serre [**145**].)

THEOREM. *Let G be an algebraic group defined over a number field F, and suppose the Hasse Principle holds for G. Then all classes in $G(F)$ satisfy the Local–Global Principle if and only if the Hasse Principle holds for all centralizers $C_G(x)$, $x \in G(F)$.*

Combined with Asai's theorem, this allows one to verify the Hasse Principle indirectly for centralizers of rational points in classical groups.

10.9. Rational Elements in Conjugacy Classes

If G is defined over a subfield k of K, we get a natural map from the rational points $G(k)$ to the collection of conjugacy classes defined over k. When is this map injective or surjective? When k is finite, we saw in 8.4–8.5 how Lang's Theorem leads to precise answers.

In his paper on regular elements [**184**, §9–11], Steinberg considers somewhat more general fields of definition, usually assumed to be perfect. A field k is said to have **cohomological dimension** ≤ 1 if all finite-dimensional divison algebras over k are commutative (for example, k finite). This is known to imply that every connected algebraic group over k is **quasi-split**, i.e., has a Borel subgroup defined over k.

He begins by investigating the conditions under which his cross-section $\mathcal{C}$ of regular classes (4.15) can be constructed over k. For this it is necessary that G be quasi-split: the unique unipotent element in $\mathcal{C}$ (representing the regular unipotent class) will lie in $G(k)$, forcing the unique Borel subgroup containing it to be defined over k. In the other direction, he shows that when G is quasi-split and has no simple factor of type A_r (r even), $\mathcal{C}$ can in fact be constructed over k. (A weaker statement is proved in the exceptional case.) As a consequence, every regular class defined over k must meet $G(k)$. This leads in turn to information about semisimple classes in the quasi-split case:

THEOREM. *Let G be simply connected and semisimple, defined over k. Then*

(i) *G is quasi-split over k if and only if every semisimple class defined over k contains an element of $G(k)$.*

(ii) *When* $\dim k \leq 1$, *G is quasi-split over k and the natural map from semisimple elements over k to semisimple classes over k is bijective.*

Formulated in terms of Galois cohomology, such results yield vanishing theorems of the type conjectured earlier by Serre.

In later work, Kottwitz [**99**] needs to allow G to be reductive but not necessarily semisimple. Adapting Steinberg's results to this situation is not as straightforward as one might expect. He is able to prove: *Let G be a connected reductive group, defined over a perfect field k of characteristic* 0 *or sufficiently large prime characteristic. Assume that G is quasi-split over k and has a simply connected derived group. Then each conjugacy class of G defined over k meets $G(k)$. For semisimple classes, no restriction on the characteristic is needed.*

The proof is organized around separate treatments of semisimple classes and of unipotent classes. For the latter, use is made of the Dynkin–Kostant theory in characteristic 0, which is adaptable to sufficiently large prime characteristics (see Chapter 7 above).

An underlying concern of Kottwitz is the case of reductive groups for which the derived group fails to be simply connected. Here the theorem can fail, but he is able to analyze the extent of the failure in terms of Galois cohomology. Along the way he introduces a notion of *stable conjugacy* for elements of $G(k)$, which is intermediate between conjugacy in $G(k)$ and conjugacy in $G(K)$. All of this is motivated by questions arising in the Langlands program.

References

1. W.A. Adkins and S.H. Weintraub, *Algebra*, Springer-Verlag, New York, 1992.
2. A.V. Alekseevskii, *Component groups of centralizers of unipotent elements in semisimple algebraic groups*, Akad. Nauk Gruzin SSR Trudy Tbiliss. Mat. Inst. Razmadze **62** (1979), 5–27. (MR 81k:20063)
3. D. Alvis and G. Lusztig, *On Springer's correspondence for simple groups of type $E_n(n = 6, 7, 8)$* (with an appendix by N. Spaltenstein), Math. Proc. Cambridge Philos. Soc. **92** (1982), 65–78.
4. T. Asai, *The conjugacy classes in the unitary, symplectic and orthogonal groups over an algebraic number field*, J. Math. Kyoto Univ. **16** (1976), 325–350.
5. P. Bala and R.W. Carter, *The classification of unipotent and nilpotent elements*, Indag. Math. **36** (1974), 94–97.
6. ______, *Classes of unipotent elements in simple algebraic groups*, Math. Proc. Cambridge Philos. Soc., I, **79** (1976), 401–425; II, **80** (1976), 1–18.
7. P. Bardsley and R.W. Richardson, *Étale slices for algebraic transformation groups in characteristic p*, Proc. London Math. Soc. **51** (1985), 295–317.
8. H.-J. Bartels, *Zur Arithmetik von Konjugationsklassen in algebraischen Gruppen*, J. Algebra **70** (1981), 179–199.
9. W.M. Beynon and G. Lusztig, *Some numerical results on the characters of exceptional Weyl groups*, Math. Proc. Cambridge Philos. Soc. **84** (1978), 417–426.
10. W.M. Beynon and N. Spaltenstein, *The computation of Green functions of finite Chevalley groups of type E_n $(n = 6, 7, 8)$*, Report No. 23, U. Warwick Computer Centre, 1982.
11. ______, *Green functions of finite Chevalley groups of type E_n $(n = 6, 7, 8)$*, J. Algebra **88** (1984), 584–614.
12. A. Borel, *Sur la cohomologie des espaces fibrés principaux et des espaces homogènes des groupes de Lie compacts*, Ann. of Math. **57** (1953), 115–207.
13. ______, *Topics in the homology theory of fibre bundles*, Lecture Notes in Math., vol. 36, Berlin, Springer-Verlag, 1967.

14. ———, *Properties and linear representations of Chevalley groups*, pp. 1–55, *Seminar on Algebraic Groups and Related Finite Groups*, Lecture Notes in Math., vol. 131, Springer-Verlag, Berlin, 1970.
15. ———, *Linear Algebraic Groups*, 2nd enlarged ed., Springer-Verlag, New York, 1991.
16. A. Borel and J. de Siebenthal, *Les sous-groupes fermés de rang maximum des groupes de Lie clos*, Comment. Math. Helv. **23** (1949/50), 200–221.
17. A. Borel and J. Tits, *Groupes réductifs*, Inst. Hautes Études Sci. Publ. Math. **27** (1965), 55–151.
18. W. Borho, J.-L. Brylinski, and R. MacPherson, *Springer's Weyl group representations through characteristic classes of cone bundles*, Math. Ann. **278** (1987), 273–289.
19. ———, *Nilpotent Orbits, Primitive Ideals, and Characteristic Classes*, Progress in Math., vol. 78, Birkhäuser, Boston, 1989.
20. W. Borho and H. Kraft, *Über Bahnen und deren Deformationen bei linearen Aktionen reduktiver Gruppen*, Comment. Math. Helv. **54** (1979), 61–104.
21. W. Borho and R. MacPherson, *Représentations des groupes de Weyl et homologie d'intersection pour les variétés nilpotentes*, C.R. Acad. Sci. Paris **292** (1981), 707–710.
22. ———, *Partial resolutions of nilpotent varieties*, pp. 23–74, *Analyse et topologie sur les espaces singuliers*, Astérisque **101–102** (1983).
23. N. Bourbaki, *Groupes et algèbres de Lie*, Chapters 4–6, Hermann, Paris, 1968; 2nd ed., Masson, Paris, 1981.
24. ———, *Groupes et algèbres de Lie*, Chapters 7–8, Hermann, Paris, 1975.
25. N. Burgoyne and C. Williamson, *Semi-simple classes in Chevalley type groups*, Pacific J. Math. **70** (1977), 83–100.
26. R.W. Carter, *Simple Groups of Lie Type*, Wiley–Interscience, New York, 1972.
27. ———, *Centralizers of semisimple elements in finite groups of Lie type*, Proc. London Math. Soc. **37** (1978), 491–507.
28. ———, *Centralizers of semisimple elements in the finite classical groups* Proc. London Math. Soc. **42** (1981), 1–41.
29. ———, *Finite Groups of Lie Type: Conjugacy Classes and Complex Characters*, Wiley–Interscience, New York, 1985.
30. B. Chang, *The conjugate classes of Chevalley groups of type* (G_2), J. Algebra **9** (1968), 190–211.
31. D.H. Collingwood and W.M. McGovern, *Nilpotent Orbits in Semisimple Lie Algebras*, Van Nostrand Reinhold, New York, 1993.
32. C.W. Curtis, *Modular representations of finite groups with split (B,N)-pairs*, pp. 57–95, *Seminar on Algebraic Groups and Related Finite Groups*, Lecture Notes in Math., vol. 131, Springer-Verlag, Berlin, 1970.
33. C. De Concini, G. Lusztig, and C. Procesi, *Homology of the zero-set of a nilpotent vector field on a flag manifold*, J. Amer. Math. Soc. **1** (1988),

15–34.

34. C. De Concini and C. Procesi, *Symmetric functions, conjugacy classes and the flag variety*, Invent. Math. **64** (1981), 203–220.
35. P. Deligne and G. Lusztig, *Representations of reductive groups over finite fields*, Ann. Math. **103** (1976), 103–161.
36. M. Demazure, *Invariants symétriques entiers des groupes de Weyl et torsion*, Invent. Math. **21** (1973), 287–301.
37. D.I. Deriziotis, *The Brauer complex of a Chevalley group*, J. Algebra **70** (1981), 261–269.
38. ______, *Centralizers of semisimple elements in a Chevalley group*, Comm. Algebra **9** (1981), 1997–2014.
39. ______, *The centralizers of semisimple elements of the Chevalley groups E_7 and E_8*, Tokyo J. Math. **6** (1983), 191–216.
40. ______, *Conjugacy Classes and Centralizers of Semisimple Elements in Finite Groups of Lie Type*, Vorlesungen aus dem Fachbereich Mathematik der Universität Essen, No. 11, 1984.
41. ______, *On the number of conjugacy classes in finite groups of Lie type*, Comm. Algebra **13** (1985), 1019–1045.
42. D.I. Deriziotis and A.P. Fakiolas, *The maximal tori in the finite Chevalley groups of type E_6, E_7 and E_8*, Comm. Algebra **19** (1991), 889–903.
43. D.I. Deriziotis and D.F. Holt, *The Möbius function of the lattice of closed subsystems of a root system*, Comm. Algebra **21** (1993), 1543–1570.
44. D.I. Deriziotis and M.W. Liebeck, *Centralizers of semisimple elements in finite twisted groups of Lie type*, J. London Math. Soc. **31** (1985), 48–54.
45. D.I. Deriziotis and G.O. Michler, *Character tables and blocks of finite simple triality groups ${}^3\mathrm{D}_4(q)$*, Trans. Amer. Math. Soc. **303** (1987), 39–70.
46. F. Digne and J. Michel, *Representations of Finite Groups of Lie Type*, London Math. Soc. Student Texts, No. 21, Cambridge Univ. Press, Cambridge, 1991.
47. S. Donkin, *On conjugating representations and adjoint representations of semisimple groups*, Invent. Math. **91** (1988), 137–145.
48. ______, *The normality of closures of conjugacy classes of matrices*, Invent. Math. **101** (1990), 717–736.
49. E. Dynkin, *Semisimple subalgebras of semisimple Lie algebras*, Amer. Math. Soc. Transl. (Ser. 2) **6** (1957), 111–245.
50. A.G. Elašvili, *Centralizers of nilpotent elements in semisimple Lie algebras*, Sakharth. SSR Mecn. Akad. Math. Inst. Šrom. **46** (1975), 109–132. (MR 52 #13958)
51. G.B. Elkington, *Centralizers of unipotent elements in semisimple algebraic groups*, J. Algebra **23** (1972), 137–163.
52. H. Enomoto, *The conjugacy classes of Chevalley groups of type (G_2) over finite fields of characteristic* 2 *or* 3, J. Fac. Sci. Univ. Tokyo **16** (1970), 497–512.

53. W. Ferrer Santos, *Closed conjugacy classes, closed orbits and structure of algebraic groups*, Comm. Algebra **19** (1991), 3241–3248.
54. P. Fleischmann and I. Janiszczak, *The lattices and Möbius functions of stable closed subrootsystems and hyperplane complements for classical Weyl groups*, Manuscripta Math. **72** (1991), 375–403.
55. ______, *The number of regular semisimple elements for Chevalley groups of classical type*, J. Algebra **155** (1993), 482–528.
56. ______, *The semisimple conjugacy classes of finite groups of Lie type E_6 and E_7*, Comm. Algebra **21** (1993), 93–161.
57. ______, *The semisimple conjugacy classes and the generic class number of the finite simple groups of Lie type E_8*, Comm. Algebra **22** (1994), 2221–2303.
58. M. Gauger, *Conjugacy in a semisimple Lie algebra is determined by similarity under fundamental representations*, J. Algebra **48** (1977), 382–389.
59. M. Gerstenhaber, *On dominance and varieties of commuting matrices*, Ann. of Math. **73** (1961), 324–348.
60. ______, *Dominance over the classical groups*, Ann. of Math. **74** (1961), 532–569.
61. V. Ginsburg, *$\mathfrak{g}$-modules, Springer's representations and bivariant Chern classes*, Adv. Math. **61** (1986), 1–48.
62. J.A. Green, *The characters of the finite general linear groups*, Trans. Amer. Math. Soc. **80** (1955), 402–447.
63. J.A. Green, G.I. Lehrer, and G. Lusztig, *On the degrees of certain group characters*, Quart. J. Math. Oxford Ser. (2) **27** (1976), 1–4.
64. A. Grothendieck, *Éléments de géométrie algébrique* III, Inst. Hautes Études Sci. Publ. Math. **11** (1961).
65. ______, *Éléments de géométrie algébrique* IV (seconde partie), Inst. Hautes Études Sci. Publ. Math. **24** (1965).
66. R. Hartshorne, *Algebraic Geometry*, Springer-Verlag, New York, 1977.
67. R.A. Herb and N.R. O'Brian, *A characterization of unipotent, semisimple and regular elements in a reductive algebraic group*, Bull. London Math. Soc. **8** (1976), 233–238.
68. W.H. Hesselink, *Singularities in the nilpotent scheme of a classical group*, Trans. Amer. Math. Soc. **222** (1976), 1–32.
69. ______, *Cohomology and the resolution of the nilpotent variety*, Math. Ann. **225** (1976), 249–252.
70. ______, *Nilpotency in classical Lie algebras over a field of characteristic* 2, Math. Z. **166** (1979), 165–18.
71. ______, *The normality of closures of orbits in a Lie algebra*, Comment. Math. Helv. **54** (1979), 105–110.
72. G. Hiss, *Die adjungierten Darstellungen der Chevalley-Gruppen*, Arch. Math. **42** (1984), 408–416.

73. G.M.D. Hogeweij, *Almost-classical Lie algebras*, Indag. Math. **44** (1982), I. 441–452, II. 453–460.
74. D.F. Holt and N. Spaltenstein, *Nilpotent orbits of exceptional Lie algebras over algebraically closed fields of bad characteristic*, J. Austral. Math. Soc. Ser. A **38** (1985), 330–350.
75. R. Hotta, *On Springer's representations*, J. Fac. Sci. Univ. Tokyo **28** (1982), 863–876.
76. ______, *On Joseph's construction of Weyl group representations*, Tôhoku Math. J. **36** (1984), 49–74.
77. ______, *A local formula for Springer's representation*, pp. 127–139, *Algebraic groups and related topics* (R. Hotta, ed.), Adv. Stud. Pure Math., vol. 6, North-Holland, Amsterdam, 1985.
78. R. Hotta and M. Kashiwara, *The invariant holonomic system on a semisimple Lie algebra*, Invent. Math. **75** (1984), 327–358.
79. R. Hotta and N. Shimomura, *The fixed point subvarieties of unipotent transformations on generalized flag varieties and the Green functions*, Math. Ann. **241** (1979), 193–208.
80. R. Hotta and T.A. Springer, *A specialization theorem for certain Weyl group representations and an application to the Green polynomials of unitary groups*, Invent. Math. **41** (1977), 113–127.
81. J.E. Humphreys, *Algebraic groups and modular Lie algebras*, Mem. Amer. Math. Soc., No. 71, Amer. Math. Soc., 1967.
82. ______, *Introduction to Lie Algebras and Representation Theory*, Springer-Verlag, New York, 1972.
83. ______, *Representations of* $\mathrm{SL}(2,p)$, Amer. Math. Monthly **82** (1975), 21–39.
84. ______, *Linear Algebraic Groups*, Springer-Verlag, New York, 1975.
85. ______, *The Steinberg representation*, Bull. Amer. Math. Soc. **16** (1987), 247–263.
86. ______, *Reflection Groups and Coxeter Groups*, Cambridge Univ. Press, Cambridge, 1990.
87. J.C. Jantzen, *Moduln mit einem höchsten Gewicht*, Lecture Notes in Math., vol. 750, Springer-Verlag, Berlin, 1979.
88. ______, *Einhüllende Algebren halbeinfacher Lie-Algebren*, Springer-Verlag, Berlin, 1983.
89. ______, *Representations of Algebraic Groups*, Academic Press, Orlando, 1987.
90. D.S. Johnston and R.W. Richardson, *Conjugacy classes in parabolic subgroups of semisimple algebraic groups*, II, Bull. London Math. Soc. **9** (1977), 245–250.
91. A. Joseph, *On the variety of a highest weight module*, J. Algebra **88** (1984), 238–278.

92. M. Kashiwara and T. Tanisaki, *The characteristic cycles of holonomic systems on a flag manifold*, Invent. Math. **77** (1984), 185–198.
93. D. Kazhdan, *Proof of Springer's hypothesis*, Israel J. Math. **28** (1977), 272–286.
94. D. Kazhdan and G. Lusztig, *Representations of Coxeter groups and Hecke algebras*, Invent. Math. **53** (1979), 165–184.
95. ______, *A topological approach to Springer's representations*, Adv. Math. **38** (1980), 222–228.
96. ______, *Fixed point varieties on affine flag manifolds*, Israel J. Math. **62** (1988), 129–168.
97. B. Kostant, *The principal three-dimensional subgroup and the Betti numbers of a complex simple Lie group*, Amer. J. Math. **81** (1959), 973–1032.
98. ______, *Lie group representations on polynomial rings*, Amer. J. Math. **85** (1963), 327–404.
99. R.E. Kottwitz, *Rational conjugacy classes in reductive groups*, Duke Math. J. **49** (1982), 785–806.
100. H. Kraft, *Parametrisierung von Konjugationsklassen in* $\mathfrak{sl}_n$, Math. Ann. **234** (1978), 209–220.
101. ______, *Closures of conjugacy classes in* G_2, J. Algebra **126** (1989), 454–465.
102. H. Kraft and C. Procesi, *Closures of conjugacy classes of matrices are normal*, Invent. Math. **53** (1979), 227–247.
103. ______, *Minimal singularities in* GL_n, Invent. Math. **62** (1981), 503–515.
104. ______, *On the geometry of conjugacy classes in classical groups*, Comment. Math. Helv. **57** (1982), 539–602.
105. J.F. Kurtzke, *Centralizers of irregular elements in reductive algebraic groups*, Pacific J. Math. **104** (1983), 133–154.
106. ______, *Centers of centralizers in reductive algebraic groups*, Comm. Algebra **19** (1991), 3393–3410.
107. S. Lang, *Algebraic groups over finite fields*, Amer. J. Math. **78** (1956), 555–563.
108. G.I. Lehrer, *Rational tori, semisimple orbits and the topology of hyperplane complements*, Comment. Math. Helv. **67** (1992), 226–251.
109. G.I. Lehrer and T. Shoji, *On flag varieties, hyperplane complements and Springer representations of Weyl groups*, J. Austral. Math. Soc. Ser. A **49** (1990), 449–485.
110. T. Levasseur and S.P. Smith, *Primitive ideals and nilpotent orbits in type* G_2, J. Algebra **114** (1988), 81–105.
111. P. Lorist, *The geometry of* $\mathcal{B}_x$, Indag. Math. **48** (1986), 423–442.
112. B. Lou, *The centralizer of a regular unipotent element in a semisimple algebraic group*, Bull. Amer. Math. Soc. **74** (1968), 1144–1177.
113. G. Lusztig, *On the Green polynomials of classical groups*, Proc. London Math. Soc. **33** (1976), 443–475.

114. ______, *On the finiteness of the number of unipotent classes*, Invent. Math. **34** (1976), 201–213.
115. ______, *Irreducible representations of finite classical groups*, Invent. Math. **43** (1977), 125–175.
116. ______, *A class of irreducible representations of a Weyl group*, Indag. Math. **41** (1979), 323–335.
117. ______, *Green polynomials and singularities of unipotent classes*, Adv. Math. **42** (1981), 169–178.
118. ______, *A class of irreducible representations of a Weyl group*, II, Indag. Math. **44** (1982), 219–226.
119. ______, *Intersection cohomology complexes on a reductive group*, Invent. Math. **75** (1984), 205–272.
120. ______, *On the character values of finite Chevalley groups at unipotent elements*, J. Algebra **104** (1986), 146–194.
121. ______, *Cells in affine Weyl groups*, IV, J. Fac. Sci. Univ. Tokyo **36** (1989), 297–328.
122. ______, *Green functions and character sheaves*, Ann. Math. **131** (1990), 355–408.
123. G. Lusztig and N. Spaltenstein, *Induced unipotent classes*, J. London Math. Soc. **19** (1979), 41–52.
124. ______, *On the generalized Springer correspondence for classical groups*, pp. 289–316, *Algebraic groups and related topics* (R. Hotta, ed.), Adv. Stud. Pure Math., vol. 6, North-Holland, Amsterdam, 1985.
125. I.G. Macdonald, *Symmetric Functions and Hall Polynomials*, Clarendon Press, Oxford, 1979.
126. G. Malle, *Die unipotenten Charaktere von* ${}^2F_4(q^2)$, Comm. Algebra **18** (1990), 2361–2381.
127. ______, *Green functions for groups of types* E_6 *and* F_4 *in characteristic* 2, Comm. Algebra **21** (1993), 747–798.
128. V.B. Mehta and W. van der Kallen, *A simultaneous Frobenius splitting for closures of conjugacy classes of nilpotent matrices*, Compositio Math. **84** (1992), 211–221.
129. ______, *On a Grauert–Riemenschneider vanishing theorem for Frobenius split varieties in characteristic* p, Invent. Math. **108** (1992), 11–13.
130. K. Mizuno, *The conjugate classes of Chevalley groups of type* E_6, J. Fac. Sci. Univ. Tokyo **24** (1977), 525–563.
131. ______, *The conjugate classes of unipotent elements of the Chevalley groups* E_7 *and* E_8, Tokyo J. Math. **3** (1980), 391–461.
132. D. Mumford, *Algebraic Geometry I: Complex Algebraic Varieties*, Springer-Verlag, Berlin, 1976.
133. ______, *The Red Book of Varieties and Schemes*, Lecture Notes in Math., vol. 1358, Springer-Verlag, Berlin, 1988.

134. K. Pommerening, *Über die unipotenten Klassen reduktiver Gruppen*, J. Algebra **49** (1977), 525–536.
135. ———, *Über die unipotenten Klassen reduktiver Gruppen*, II, J. Algebra **65** (1980), 373–398.
136. R.W. Richardson, Jr., *Conjugacy classes in Lie algebras and algebraic groups*, Ann. Math. **86** (1967), 1–15.
137. ———, *Conjugacy classes in parabolic subgroups of semisimple algebraic groups*, Bull. London Math. Soc. **6** (1974), 21–24.
138. ———, *The conjugating representation of a semisimple group*, Invent. Math. **54** (1979), 229–245.
139. ———, *Derivatives of invariant polynomials on a semisimple Lie algebra*, pp. 228–241, *Harmonic analysis and operator theory*, Proc. Cent. Math. Anal. Austral. Natl. Univ. **15**, 1987.
140. ———, *Conjugacy classes of n-tuples in Lie algebras and algebraic groups*, Duke Math. J. **57** (1988), 1–35.
141. R. Richardson, G. Röhrle, and R. Steinberg, *Parabolic subgroups with Abelian unipotent radical*, Invent. Math. **110** (1992), 649–671.
142. G.B. Seligman, *Some remarks on classical Lie algebras*, J. Math. Mech. **6** (1957), 549–558.
143. ———, *Modular Lie Algebras*, Springer-Verlag, Berlin, 1967.
144. J.-P. Serre, *Groupes algébriques et corps de classes*, Hermann, Paris, 1959; transl. *Algebraic Groups and Class Fields*, Springer-Verlag, New York, 1988.
145. ———, *Cohomologie galoisienne: progrès et problèmes*, Sem. Bourbaki (1993–94), Exposé 783, Astérisque.
146. N. Shimomura, *A theorem on the fixed point set of a unipotent transformation on the flag manifold*, J. Math. Soc. Japan **32** (1980), 55–64.
147. K. Shinoda, *The conjugacy classes of Chevalley groups of type* (F_4) *over finite fields of characteristic* 2, J. Fac. Sci. Univ. Tokyo **21** (1974), 133–159.
148. ———, *The conjugacy classes of the finite Ree groups of type* (F_4), J. Fac. Sci. Univ. Tokyo **22** (1975), 1–15.
149. T. Shoji, *The conjugacy classes of Chevalley groups of type* (F_4) *over finite fields of characteristic* $p \neq 2$, J. Fac. Sci. Univ. Tokyo **21** (1974), 1–17.
150. ———, *On the Springer representations of the Weyl groups of classical algebraic groups*, Comm. Algebra **7** (1979), 1713–1745; correction, ibid., 2027–2033.
151. ———, *On the Springer representations of Chevalley groups of type* F_4, Comm. Algebra **8** (1980), 409–440.
152. ———, *On the Green polynomials of a Chevalley group of type* F_4, Comm. Algebra **10** (1982), 505–543.
153. ———, *On the Green polynomials of classical groups*, Invent. Math. **74** (1983), 239–267.
154. ———, *Geometry of orbits and Springer correspondence*, pp. 61–140, *Orbites unipotentes et représentations*, I, Astérisque **168** (1988).

155. P. Slodowy, *Four lectures on simple groups and singularities*, Comm. Math. Inst. Utrecht, 1980.
156. ———, *Simple Singularities and Simple Algebraic Groups*, Lecture Notes in Math., vol. 815, Springer-Verlag, Berlin, 1980.
157. N. Spaltenstein, *The fixed point set of a unipotent transformation on the flag manifold*, Indag. Math. **38** (1976), 452–456.
158. ———, *On the fixed point set of a unipotent element on the variety of Borel subgroups*, Topology **16** (1977), 203–204.
159. ———, *Nilpotent classes and sheets of Lie algebras in bad characteristic*, Math. Z. **181** (1982), 31–48.
160. ———, *Classes unipotentes et sous-groupes de Borel*, Lecture Notes in Math., vol. 946, Springer-Verlag, Berlin, 1982.
161. ———, *On unipotent and nilpotent elements of groups of type E_6*, J. London Math. Soc. **27** (1983), 413–420.
162. ———, *A property of special representations of Weyl groups*, J. Reine Angew. Math. **343** (1983), 212–220.
163. ———, *Nilpotent classes in Lie algebras of type F_4 over fields of characteristic* 2, J. Fac. Sci. Univ. Tokyo **30** (1984), 517–524.
164. ———, *Existence of good transversal slices to nilpotent orbits in good characteristics*, J. Fac. Sci. Univ. Tokyo **31** (1984), 283–286.
165. ———, *On the generalized Springer correspondence for exceptional groups*, pp. 317–338, *Algebraic groups and related topics* (R. Hotta, ed.), Adv. Stud. Pure Math., vol. 6, North–Holland, Amsterdam, 1985.
166. ———, *On Springer's representations of Weyl groups containing* -1, Arch. Math. **44** (1985), 26–28.
167. ———, *Polynomials over local fields, nilpotent orbits and conjugacy classes in Weyl groups*, pp. 191–217, *Orbites unipotentes et représentations*, I, Astérisque **168** (1988).
168. ———, *On the reflection representation in Springer's theory*, Comment. Math. Helv. **66** (1991), 618–636.
169. ———, *Order relations on conjugacy classes and the Kazhdan–Lusztig map*, Math. Ann. **292** (1992), 281–318.
170. T.A. Springer, *Some arithmetical results on semi-simple Lie algebras*, Inst. Hautes Études Sci. Publ. Math. **30** (1966), 115–142.
171. ———, *A note on centralizers in semi-simple groups*, Indag. Math. **28** (1966), 75–77.
172. ———, *The unipotent variety of a semisimple group*, pp. 373–391, *Algebraic Geometry* (ed. S. Abhyankar), Oxford University Press, London, 1969.
173. ———, *Regular elements of finite reflection groups*, Invent. Math. **25** (1974), 159–198.
174. ———, *Trigonometric sums, Green functions of finite groups and representations of Weyl groups*, Invent. Math. **36** (1976), 173–207.

175. ______, *A construction of representations of Weyl groups*, Invent. Math. **44** (1978), 279–293.
176. ______, *The Steinberg function of a finite Lie algebra*, Invent. Math. **58** (1980), 211–215.
177. ______, *Linear Algebraic Groups*, Birkhäuser, Boston, 1981.
178. ______, *A purity result for fixed point varieties in flag manifolds*, J. Fac. Sci. Univ. Tokyo **31** (1984), 271–282.
179. ______, *On representations of Weyl groups*, pp. 517–536, *Proc. Hyderabad Conference on Algebraic Groups*, Manoj Prakashan, Madras, 1991.
180. T.A. Springer and R. Steinberg, *Conjugacy classes*, pp. 167–266, *Seminar on Algebraic Groups and Related Finite Groups*, Lecture Notes in Math., vol. 131, Springer-Verlag, Berlin, 1970.
181. B. Srinivasan, *Green polynomials of finite classical groups*, Comm. Algebra **5** (1977), 1241–1258.
182. R. Steinberg, *A geometric approach to the representations of the full linear group over a Galois field*, Trans. Amer. Math. Soc. **71** (1951), 274–282.
183. ______, *Representations of algebraic groups*, Nagoya Math. J. **22** (1963), 33–56.
184. ______, *Regular elements of semisimple algebraic groups*, Inst. Hautes Études Sci. Publ. Math. **25** (1965), 49–80.
185. ______, *Classes of elements of semisimple algebraic groups*, pp. 277–284, *Proc. Intern. Congress of Mathematicians* (Moscow, 1966), Mir, Moscow, 1968.
186. ______, *Lectures on Chevalley Groups*, Yale Univ. Math. Dept., 1968.
187. ______, *Endomorphisms of linear algebraic groups*, Mem. Amer. Math. Soc., No. 80, 1968.
188. ______, *Conjugacy classes in algebraic groups* (notes by V.V. Deodhar), Lecture Notes in Math., vol. 366, Springer-Verlag, Berlin, 1974.
189. ______, *Torsion in reductive groups*, Adv. Math. **15** (1975), 63–92.
190. ______, *On the desingularization of the unipotent variety*, Invent. Math. **36** (1976), 209–224.
191. ______, *On theorems of Lie–Kolchin, Borel, and Lang*, pp. 349–354, *Contributions to Algebra*, Academic Press, New York, 1977.
192. ______, *Conjugacy in semisimple algebraic groups*, J. Algebra **55** (1978), 348–350.
193. ______, *Kleinian singularities and unipotent elements*, pp. 265–270, *The Santa Cruz Conference on Finite Groups*, Proc. Sympos. Pure Math., vol. 37, Amer. Math. Soc., Providence, RI, 1980.
194. U. Stuhler, *Unipotente und nilpotente Klassen in einfachen Gruppen und Liealgebren vom Typ G_2*, Indag. Math. **33** (1971), 365–378.
195. D.M. Testerman, *A_1-type overgroups of elements of order p in semisimple algebraic groups and the associated finite groups*, J. Algebra (to appear).

196. M. Varagnolo, *Harmonic polynomials and Springer's representation for* $\mathrm{SL}(n, \mathbb{C})$, J. Algebra **153** (1992), 444–452.
197. M. Vergne, *Polynômes de Joseph et représentation de Springer*, Ann. Sci. École Norm. Sup. **23** (1990), 543–562.
198. G.E. Wall, *On the conjugacy classes in the unitary, symplectic and orthogonal groups*, J. Austral. Math. Soc. **3** (1963), 1–62.
199. ———, *Conjugacy classes in projective and special linear groups*, Bull. Austral. Math. Soc. **22** (1980), 339–364.
200. H. Weyl, *The Classical Groups*, 2nd ed., Princeton Univ. Press, 1946.

Index

Recent Titles in This Series

(*Continued from the front of this publication*)

4 **O. F. G. Schilling,** The theory of valuations, 1950
3 **M. Marden,** Geometry of polynomials, 1949
2 **N. Jacobson,** The theory of rings, 1943
1 **J. A. Shohat and J. D. Tamarkin,** The problem of moments, 1943

ISBN 0-8218-0333-6